STEEL POTS

The History of America's Steel Combat Helmets

by
Chris Armold

1st Edition

Published by R. James Bender Publishing,
P.O. Box 23456, San Jose, California 95153
Phone: (408) 225-5777
Fax: (408) 225-4739

Printed in the United States of America

ISBN No. 0-912138-70-X

TABLE OF CONTENTS

FOREWORD

Mr. Christopher W. Armold has written a fascinating and much needed book on the evolution of the American helmet. Ever since the first cave man picked up a rock (the first missile weapon), another man has tried to protect himself from the effects of that crude, but effective, piece of armament. I can remember a cartoon strip from my youth entitled "Ally Oop." You could always tell when the cavemen went to war because they were wearing turtle shells on their heads. I doubt that was really the case with *homo erectus*, but some of the earliest military paintings in the world show Egyptian soldiers fighting enemies with readily identifiable headgear which we would term a helmet. From that point on until well into the 16th century, armor became more and more sophisticated.

Throughout the bronze age, fighting men in the area of the fertile crescent used leather and bronze helmets along with other protective headgear made from cloth. One of the earliest and best known examples of a helmet from the Mesopotamian civilization is a gold helmet worn by a royal warrior which dates from the third century BC. It would seem that the Hittites were the first people to introduce the iron helmet onto the battlefield sometime around 2000 BC.

The early Greek civilizations had rather elaborate helmets for their warriors. Judging from examples dug up from Knossos, the capital of the Minoan civilization, and Mycenae, Homer accurately described typical helmets of the time in his epic poem detailing the fall of Troy (c. 1184 BC.). The Greek helmets of the day were made of leather overlaid with bronze strips or overlapping rows of boars' tusks. The later Greek civilizations saw the advent of the Corinthian helmet which we now associate with the Greek hoplite...a helmet with fixed cheek pieces, an extension covering the nape of the neck, and a fixed bar projecting down from the brow protecting the nose. This helmet was often decorated with a high crest made from horse hair and inlayed with precious metal in the form of animals, deities, or fighting men.

The Roman civilization used numerous styles of helmets for its soldiers and other events. For instance, there is a whole series of helmets used for gladiatorial combat and for parade wear. The most recognizable helmet from the Roman era is the so-called "jockey cap," a hemispherical, slightly bulbous skull cap topped with a button or crest, with two hinged cheekpieces that tied under the chin.

With the onset of the Middle Ages, armor and helmets reached their zenith with the knights and men-at-arms. Whole books can, and have been written, on such recondite subjects as the Basnet, Barbut, Armet, and Burgonet helmets. These are but a few of the helmets worn during the period.

The entire Middle Ages is characterized by constant improvements in the protection of the head and body against sword, mace, and lance.

The 17th century sees gunpowder weapons gain in ascendancy over muscle powered ones and the decline in the amount of armor worn by soldiers. Soldiers still wanted to protect their "brain housing group" and various headgear made of felt, fur, leather and brass appeared on the battlefield to do just that.

Despite the advent of gunpowder, helmets and breastplates lasted well into the 19th century on the battlefield. During the Franco-Prussian War in 1870, one can still see the vestiges of the knights in the paintings of the Battle of Seden. German heavy cavalry was still armed with a sword and wore helmets and breastplates. Never mind the fact that the French mowed them down in vast numbers with an early machine gun…the *mitrailleuse*.

In spite of the long history of armoring the head and upper body, it is an old fact that, except for the German Army, most World War I armies had completely done away with the helmet. As trench fighting became more and more prevalent it was recognized by soldiers almost immediately that casualties could be reduced significantly with the reintroduction of the helmet. Airburst artillery and machine gun fire along with the rain and muck made the trenches a living hell. A helmet made life simpler and a lot more safe. Every nation seemed to pick its own distinctive style. It is of some historic interest that our British cousins settled on a "tin hat" that strongly resembles the "shapewe" worn by English archers and infantrymen during the 15th and 16th centuries.

During the Vietnam War, soldiers and Marines complained about the World War II style helmet being hot in the sun, cold during the Monsoon season and wet when it rained. Since it had a high center of gravity it flopped around and was uncomfortable in all seasons. You could cook in it. Helmet stew was a favorite. This was nothing more than the contents of several C-Ration boxes being mixed and heated (usually with C-4 plastic explosive) and then shared among one's messmates. Upon occasion you could dig with it, bail water out of a landing craft, barf in it, and shave and bathe in it all in the same day. When you rigged it for parachuting it became like putting your head in a vice. In short it was essential. As it was essential, soldiers and Marines both hated and loved it.

Anyone who has worn one, or knows someone that has worn one, will appreciate this book. It tells a good story.

WILLIAM F. ATWATER, Ph.D.
DIRECTOR
U.S. ARMY ORDNANCE MUSEUM

AUTHOR'S COMMENTS

While the idea of writing a history of U.S. helmets was mine I've got to give the motivational credit to my friend and collecting mentor Don Leffel. It was Don who told me that if I'm going to write a book it had better be accurate and scholarly. Man, I sure tried! I had a very clear vision of what I wanted to do so I tried to take advantage of every opportunity available. Consequently, this may be the first book written for the militaria collector that capitalized on the power of the Internet. I created "The Steel Pots Page," a small "this is what I'm doing" homepage on the World Wide Web. The reaction to it was very positive. I worked on this book for nearly three years, and here are a few things I would like you to know. First, I've tried very hard not to disguise my own opinions as facts, consequently you shouldn't encounter any guesses in this book. Secondly, I've tried to make "Steel Pots" as complete as possible but there comes a point in writing a book where you must stop. If there are omissions or mistakes, you know who to blame! Thirdly, I'll continue to gather information about U.S. helmets so please drop me a letter or an e-mail if you have some interesting information or photographs. I'd love to hear your thoughts and I'll be more than happy to clarify points or issues in the book. Finally, some special thanks: My wife Connie for her support and encouragement during my research, photography and writing. My Mom for just believing in my abilities to do whatever I put my mind to. And my friend Bernie Johnson, whom I miss terribly. Finally, Roger Bender, his wife, and the staff at R. James Bender Publishing for having the faith and trust to make "Steel Pots, The History of America's Steel Combat Helmets" a reality.

Chris Armold, 6 June 1997
chrisa@aristotle.net

For My Three Favorite Collectors:

Jerry Nichols, Jr.
John Gibson
Don Leffel

ACKNOWLEDGEMENTS

Agiorgousis, Evangelos
Andrews, John
Armold, Joanne
Armold, Joe
Armold, Nick
Armold, Tim
Atwater, William
Bender, R. James
Blaisdel, Hal
Blakeman, Chuck
Bradford, Walter
Brown, Bill
Charbonnier, Philippe
Coy, John
Deguisne, Dan
DeLaurant, Dave
Elliott, Scott
Farrar, Scott
Findley, Jim
Garland, Ian
Gawne, Jon
Gibson, John
Godin, Roger
Gordon, Sharon
Gordon, Tom
Haley, Glenn
Hall, Mike
Hanson, Luther
Harper, Frank
Harper, Mickey
Heasley, Ed
Hetrick, Tony
Hicks, Ed
Huff, Terry
Hutton, Pat
Jackson, Tanya
Jamieson, Keith
Johnson, Bernie
Krane, Benjamin
Kumer, Richard
LaRocca, Donald
Law, Clive
Lee, Jerry
Leffel, Ben
Leffel, Don
Lehmacher, Bob
Leinicke, Kris
Lewis, Kenneth
Lindsley, Donald
Long, Barry
Lucy, Roger
Martin, Frank
Mendyke, Don
Mendyke, Donna
Munnikhuysen III, Lawrence
Murphy, Joann
Nichols, Jerry
Novello Jr., Marcello
O'Dell, Susan
O'Donnell, Patrick
Orwan, Michael
Pahl, Al
Petersen, George A.
Phillips, Jim
Powers, David
Rengers, Carrie
Reynosa, Mark
Rizzutti, Bill
Simpson, Chuck
Sloat, Tim
Slonaker, Robert
Smith, Col. Barry
Smith, Scott
Smith-Christmas, K.L.
Sommala, Prinya
Southee, Robert
Spangler, John
Sregzinski, Lois
Steen, Shawndale
Stratton, Skip
Sweeden, Edward
Tagliavini, Michele
Wade, Melody
Waiss, Chris
Weiblen, Jerry
West, Mac
Wolcott, Stan

Special thanks to all the people on the Internet who supported *"The Steel Pots Page," "The Militaria Collector's Exchange"* and the advanced helmet collectors on *HelmNet.*

Aristotle Internet Access
Metropolitan Museum of Art
National Archives of the United States of America
Rock Island Arsenal Museum
The G.I. Journal
The Smithsonian Institute
U.S. Army Center For Military History
U.S. Air Force Museum
U.S. Army Military History Institute
U.S. Army Ordnance Museum
U.S. Army Quartermaster Museum
U.S. Army Research Laboratory
U.S.M.C. Air and Ground Museum
Wal-mart Store #24
Westinghouse Electric Corporation
314th Airlift Wing Public Affairs
314th Contracting Squadron
314th Security Police Squadron

All styrofoam heads used in *Steel Pots* courtesy of:
Ricky's Gun Room, Inc.
P.O. Box 286
Chicopee, MA 01021-0286

CHAPTER ONE

THE BADGE OF THE ENGLISH-SPEAKING SOLDIER

In 1916 with America's entry into the war in Europe looming ever closer, American military planners embarked on a journey to equip the men who would serve their country. In the eyes of many, protective steel helmets were considered novelty items; far less important than boots, weapons and general issue clothing.

The importance of a steel protective helmet became clear when casualty reports from European battlefields began reaching U.S. shores--shrapnel from artillery and grenades accounted for a higher rate of injury and death to combatants than bullets. Furthermore, French and British medical reports indicated that a protective helmet could prevent many injuries and fatalities attributed to shrapnel.

In 1917, the American General Staff formed a helmet committee to evaluate British, French and U.S. prototype designs for possible adoption as combat headgear for U.S. soldiers.

It soon became obvious that the selection of a helmet could not wait for an in-depth analysis or evaluation. Great Britain, eager to have her allies from across the Atlantic join the war, offered approximately 400,000 British-style combat helmets immediately. That offer, coupled with test results showing the British helmet's superior resistance to penetration compared to the French style, prompted the helmet committee to recommend the British-style combat helmet.

The American General Staff approved the recommendation, giving birth to the Model 1917 steel combat helmet.

Although initially considered a stop-gap measure until a distinctive, suitable American combat helmet could be developed, the M-1917 design would endure as the standard American armed forces steel combat helmet for almost 25 years.

Immediately after approval of the helmet, the U.S. Army Ordnance Department displayed British-made examples of the helmet to U.S. manufacturers of steel components, seeking production bids.

Initial attempts at production were disappointing as the manufacturing process was far from refined. American's manufacturing forces experimented with several formulas and types of steel and initially decided to make use of chrome-nickel steel. They soon discovered that fabrication from chrome-nickel steel was a labor intensive process which required five separate metal-pressing operations, or draws, to form the helmet shape. Many times the bowl of the helmet would emerge from the steel press wrinkled or deformed. Then came the task of annealing, or heat treating, the steel helmet to strengthen it and eliminate brittleness. Annealing was a tedious task in which an incorrect temperature could destroy an entire batch. The overall difficulty of manufacturing helmets from chrome-nickel forced a search for alternatives. Another type of steel formula known as Vanadium steel was also tried but met with problems similar to the chrome-nickel steel.

While containing several subtle differences, the American-manufactured M-1917 helmet would be essentially an exact reproduction of the British helmet devised by a Mr. John Brodie in 1915. As such, American industry looked to the British Isles and opted to utilize the same steel alloy as the British. An Englishman, Sir Robert Hadfield, convinced the British government that a steel mixture with approximately 12 percent manganese steel would result in an alloy that was very strong. He believed his mixture would remain strong even after pressing and drawing the alloy into the helmet's bowl shape.

After U.S. tests, Hadfield's steel formula was accepted. Initial testing determined the steel mixture was easy to work, and most importantly, required no annealing to protect it from becoming brittle. Further tests showed the M-1917 could be pressed in a single draw from this alloy. These factors allowed for easier production and a greater uniformity of the final product. After American steel manufacturers developed the skills to provide thin rolls of the Hadfield steel, American companies went to work producing the helmets.[1]

The initial process for construction of the M-1917 helmet shell consisted of four steps. First, a steel square of Hadfield manganese steel approximately 20 inches by 20 inches was placed in a two-stage steel press. The first stage secured the steel sheet and the second stamped out the form of the helmet, in addition to four inspection indentations or punch marks.

After stamping, the now-formed helmet was inspected. During this second stage, the additional indentations punched in the corner of each sheet were inspected. If any were cracked or malformed, the helmet was rejected.

Helmets which passed the inspection advanced to stage three, where the excess steel was cut and trimmed from the helmet form. In step four, welders attached the helmet's steel rim and riveted the helmet chinstrap loops to the underside of the helmet's protective flair. The helmet was then stamped under the helmet flair with a number identifying the heat[2] of the steel.

---

[1] *Except for a single lot of 200,000, American made-helmets were constructed from Hadfield steel. (The odd-lot was produced by the Colombian Enameling Company of Terra Haute, Indiana using a nickel-manganese formula).*

[2] *The heat of the steel referred to a temperature control number which was assigned to sheets of steel for an entire lot of helmets. This was used to identify helmets in the event the steel used for a lot was discovered to be defective.*

British-American helmet in process of manufacture. The double-action presses, shown in background at the right, stamp out the helmet in a single "draw." Budd Manufacturing Co., Philadelphia.

The plate is being "blanked out" so as to form the helmet rim; in another part of the picture the thin separate metal rims are being spot-welded in place.

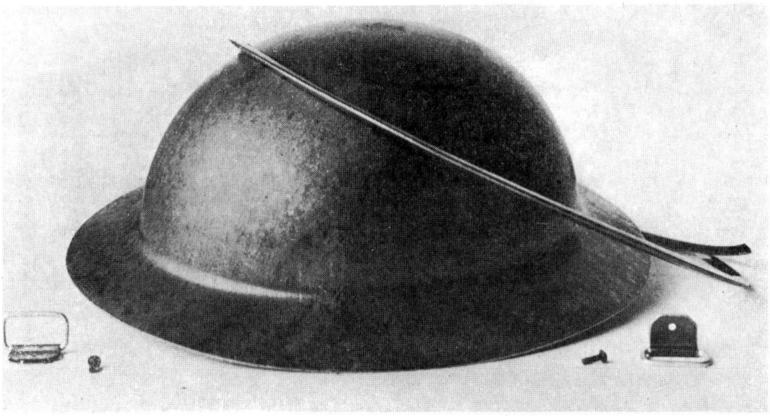

Helmet shell, metal rim, chinstrap loops and rivets ready to be put together.

Test of a plate of helmet steel. The corner of the plate is given a punch-mark; if the metal cracks, the plate is rejected.

Each lot of helmet shells was subject to test and inspection. The test consisted of firing a semi-automatic, .45 caliber, 230 grain, jacketed ball round from an American-made Colt Model 1911 pistol. Fired at a range of 10 feet, the velocity of the round was 600 feet per second.[3]

To pass this test, helmets had to resist penetration and cracks. Dents deeper than 1 3/16 inches were not permitted. Failure in any of these areas could mean rejection of an entire lot of helmets.[4]

After testing was completed, the steel shells were packaged and sent to the Ford Manufacturing Company's plant in Philadelphia, Pennsylvania to be fitted with the helmet's internal suspension.

In producing the internal suspension, the United States again decided to follow the lead of the British, instead of developing its own liner system for the M-1917. Like the British helmet, the primary components of the lining were a felt pad, a cotton twine mesh, oil cloth and leather.

Besides supporting and distributing the weight of the steel helmet shell, the liner also attempted to protect the soldier from injury if the helmet was

---

3 The ammunition used was specifically designed for this test. The standard .45 round fired from a Model 1911 traveled at approximately 830 feet per second.

4 From a lot of 200,00 helmets, a single helmet was tested from the first 50 produced. From numbers 51 to 250, two helmets were tested. From numbers 251 through 500, three helmets were checked. From 500 to 2000 shells, four were tested. 2001 to 3000 shells required the testing of six shells. After 3000 shells, the remainder of the lot required one helmet tested for every 500 produced.

struck by a projectile. Along the exterior of the lining was a leather band which ran along the perimeter of the entire liner system. This leather band was divided into an even amount of loops.[5]

Into every other loop a one-inch piece of rubber tubing was inserted. This tubing system, coupled with the suspension netting,[6] kept the steel shell away from the wearer's head so that injury from dents due to bullets or shrapnel would be minimized.

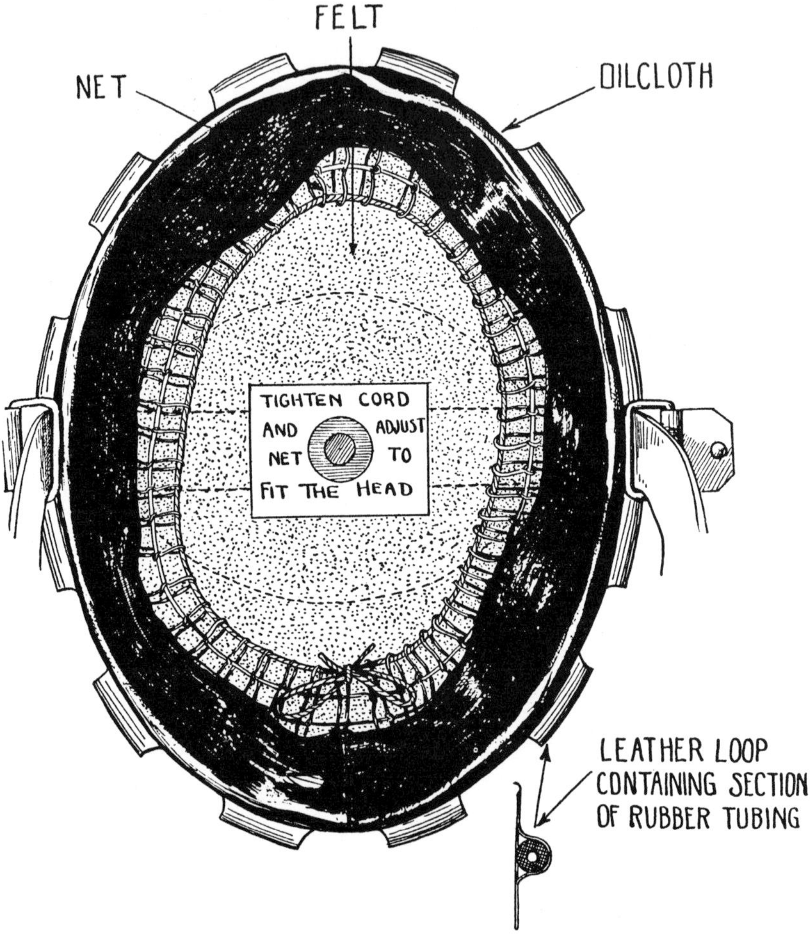

Lining of the British-American helmet.

As the helmet shells were received, they were placed 10 to a group in a special carrier to be moved through portions of the assembly and finishing process. The first stage was painting. Shells were totally immersed in a paint tank containing olive-drab paint. After dipping, the shells continued their journey to the next stage via an overhead conveyor system.

---

5 The number of loops (between 24 and 28) depended on the size of the helmet lining.

6 The netting when properly adjusted kept the steel shell of the helmet approximately one inch off the crown of the wearer's head.

Freshly painted helmets being passed along over drip-boards.

Freshly painted helmets being given a coating of sawdust (front, right).

Before they had time to dry completely, an application of fine sawdust was blown on the helmets using a compressed air system. The idea was to provide a rough texture, aiding in the prevention of glare.[7]

---

[7] *While U.S. M1917 helmets were textured using sawdust, the British-made helmets were coated with a sand mixture.*

The classic lines of the M-1917 helmet. Its exterior is painted olive drab and textured with sawdust. This helmet bears the insignia of the 91st Infantry Division.

After the sawdust application, the shells were placed in a large cage which could contain nearly 600. For the next hour, the shells endured a 200-degree oven. It cured the paint and assisted in drying.

Exiting the oven, the sawdust-covered shells were dipped in the olive-drab paint for a final coat. They then returned to the oven for another hour.

This painting and curing process complete, the helmets moved towards the final assembly stage, where their suspension system would be installed.

When the shell arrived for its liner, a workman glued a muslin-backed, 1/4-inch thick, felt oval pad into its interior crown. One half of a leather chinstrap was then secured to the leather, rubber and felt headband suspension with a single rivet on each side of the suspension. This part of the leather chinstrap with the liner attached, was then threaded through the helmet bails and positioned in the center crown of the helmet. A rivet driven through the crown of the steel shell pierced the felt pad and leather chinstrap support to hold the entire assembly in place. The second half of the chinstrap assembly was riveted to the supporting portion of the strap and the helmet assembly process was completed.

Throughout the finishing and assembly process, the helmets steadily moved from the roof of the Ford plant to the lower level. Completed helmets were placed in a special machine which compressed approximately 40 to 50 helmets together, lifted them, and placed them in a wooden crate for shipping to supply depots or overseas ports.

There were many faults with the new helmet. The helmet had a poor center of gravity and the simple action of walking caused it to bob and wobble. A very large portion of the wearer's head was left exposed. Especially from the rear, the helmet offered little protection. In combat it was a common sight to see a soldier holding the helmet on his head while running. Assuming the prone position often caused the helmet to lurch forward and either fall off or obscure vision.

In the field, damage to the internal suspension or chinstrap of the helmet could often not be repaired. The leather chinstraps were cumbersome and uncomfortable. They stretched when they were wet and shrank in hot weather.[8]

Note: British-made helmets have a large rubber "donut" ring under the felt in the crown.

The interior of the helmet contained the suspension system and leather chinstrap. Like the steel shell, the U.S. M-1917 is nearly identical to late-model British helmets of the First World War.

Even with all these faults the "tin hats" or "trench helmets" as the Doughboys called them, became a symbol of the allied forces in World War I.

[8] *An improved helmet chinstrap was developed for the M1917 helmet and was incorporated into post-war examples. Constructed of a khaki webbing and relying on a buckle and hook system, it was essentially the same chinstrap that would be used on the M1917A1 helmets.*

In fact loyalty to the British design and affection for the helmet by combat troops derailed the testing and introduction of several experimental models.

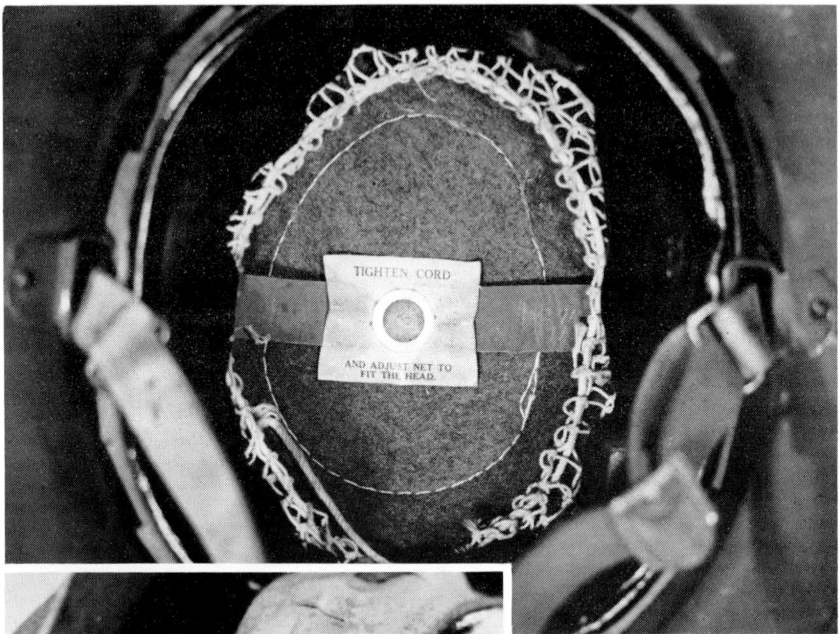

A close look at the gray oval felt pad glued to the crown of the helmet and the black oilcloth headband. Also visible is the string netting which could be adjusted to improve the helmet's fit. The paper tag in the crown reads "Tighten Cord To Adjust Net To Fit The Head." The leather strap running across the crown is part of the chinstrap.

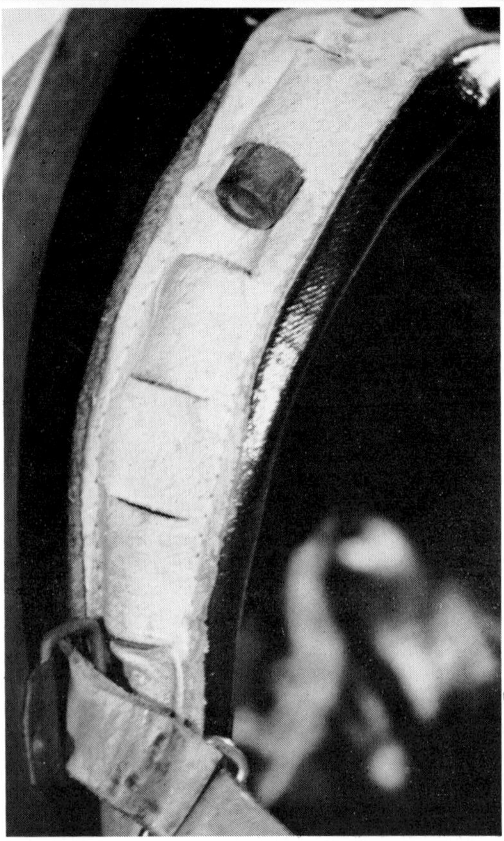

The leather perimeter of the helmet liner contains one-inch sections of rubber tubing. The tubes were used to provide padding between the helmet shell and the head, if the helmet was struck by a projectile.

The heavy two-piece chinstrap also served as the main support structure for the liner. The main rivet in the helmet crown ran through the chinstrap. A wire loop holds the two parts of the strap together. The chinstrap could be adjusted by sliding the metal buckle. The chinstrap was a weak link in the helmet's structure. If it broke or shrank due to the elements, the helmet was essentially unserviceable.

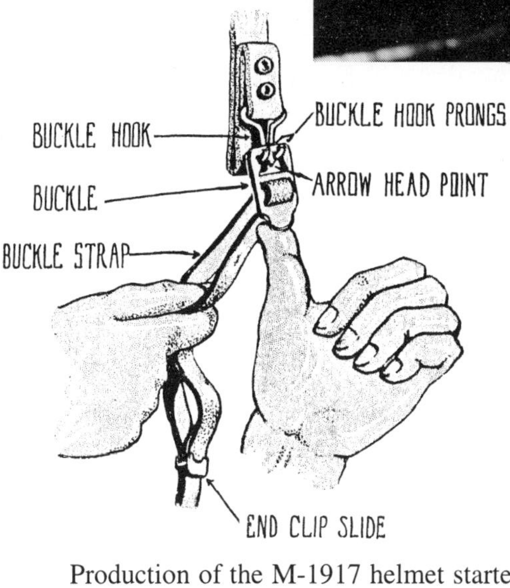

Diagram showing the method of tightening the new chinstrap; also the new buckle-hook. This late-war development was incorporated into post-war examples.

Production of the M-1917 helmet started in October 1917 in the United States. For the production of liners, 10 companies -- including the Progressive Knitting Works of New York, The Taylor Company of Buffalo, the Curtain Supply Company of Chicago, the Leatherwear Company of America operating from New York -- contributed to the war effort by producing these required items. On the steel side of the manufacturing process, the American Sheet and Tin Plate Company in Gary, Indiana assured the supply and production of the required Hadfield manganese steel for the helmet shells.

Total production of the M-1917 helmet during World War I eclipsed 2,500,000 with the following companies pressing and stamping the helmet shells:

- Benjamin Electric Company, Des Plaines, Illinois 33,600
- Bossett Corporation, Utica, New York 116,735
- Columbian Enameling and Stamping Company, Terre Haute, Indiana 268,850
- Crosby Company, Buffalo, New York 469,968
- Edward D. Budd Manufacturing Company, Philadelphia, Pennsylvania 1,150,775
- Sparks, Withington Company, Jackson, Michigan 473,469
- Worcester Pressed Steel Company, Worcester Massachusetts 193,846

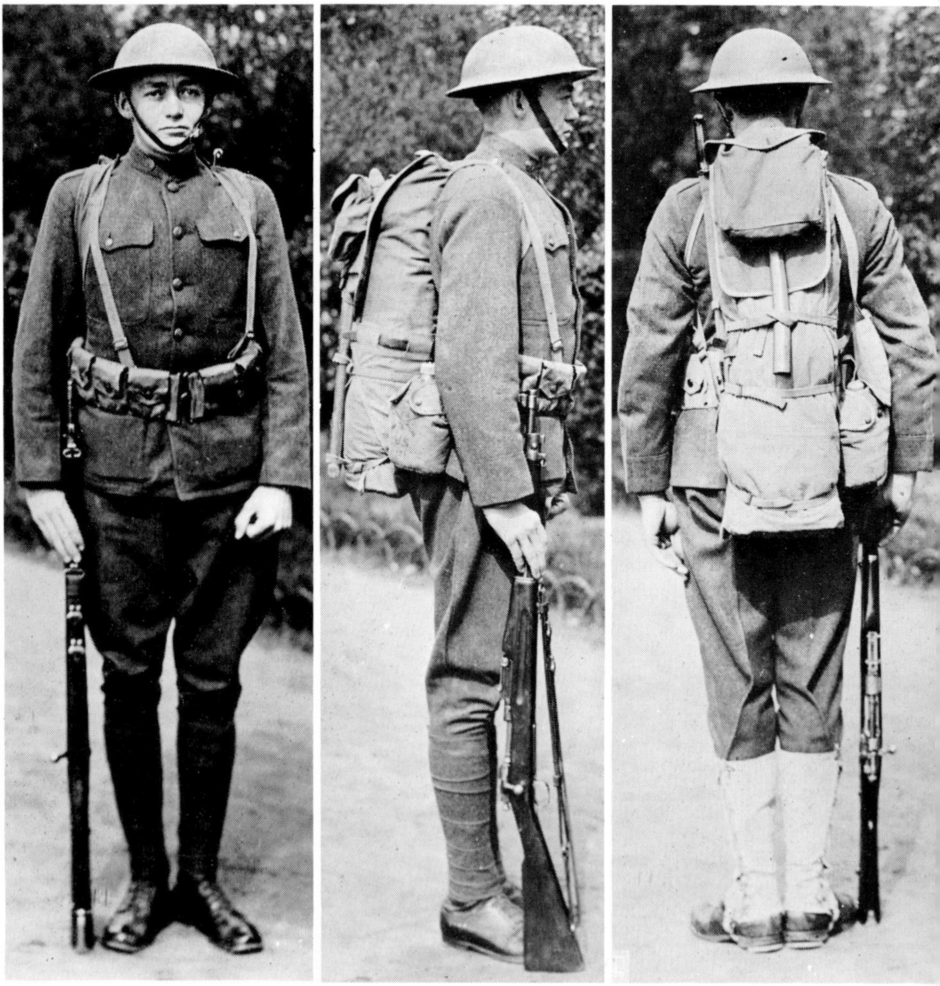

Pvt. H.T. Johnson, Company L, 28th Infantry, with helmet and full pack.

A visual difference between American-made and British-made M-1917 helmets. The American-made helmet on the left has much more texture than the relatively smooth shell of the British-made example. (Bill Brown Collection)

For the military historian or helmet collector, determining if a helmet is the British-made "Brodie's Steel Helmet" or the American-made M-1917, can be a tricky proposition. Early examples of the British-made helmet can be distinguished by:

- Adjustable buckling chinstrap
- Machine fitted round-head rivets were used to secure the chinstrap loops to the helmet. (Later versions of the Mark I used split-head rivets)
- A very shiny, smooth paint finish
- The rubber tubing used as a cushion between the steel shell and the head was wrapped in an oil cloth material rather than in leather loops.
- Chinstrap was secured directly to the chinstrap loops
- Oil cloth lining consisted of six leaves secured at the center rather than a single band around the helmet circumference.

While the differences in these early British-made helmets seem relatively straightforward, that is unfortunately not the case with helmets made after 1916. The differences between the U.S. and these later British-made helmets, now designated the Mark I, can be difficult to detect. A few clues are:

- Helmet rim often overlaps on British-made helmets.
- Chinstrap loops on British Mark I helmets manufactured after 1916 were secured to the steel shell using round-head split rivets instead of machine-fitted rivets.
- Stamps under the flair bearing the initials BS, FS, HS, or MS reflect known British steel manufacturing codes.
- Beneath the linings, the British Mark I's bear a rectangular stamp in red ink identifying the helmet as "Brodie's Steel Helmet" with patent and

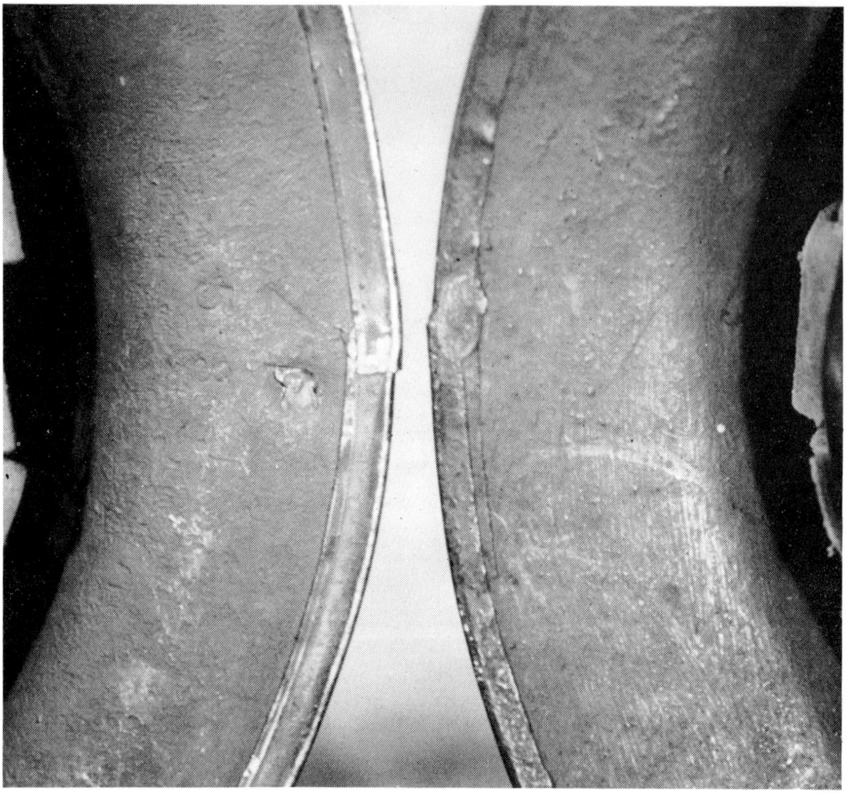

A comparison of the rim-seam welds on American (left) and British-made helmets. Typically, there is a larger overlap on British-made helmets. (Bill Brown Collection)

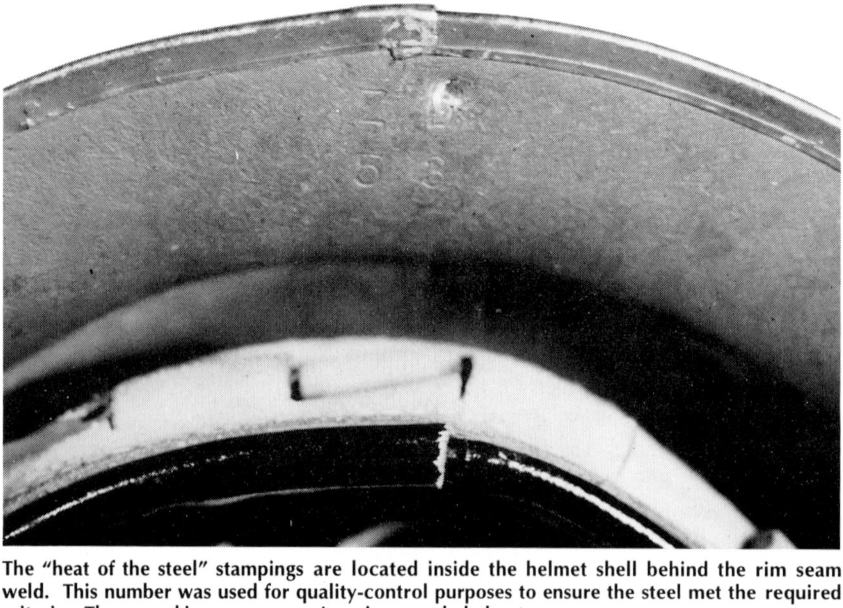

The "heat of the steel" stampings are located inside the helmet shell behind the rim seam weld. This number was used for quality-control purposes to ensure the steel met the required criteria. These markings are on an American-made helmet.

On the left, American-made and on the right, British. Notice the subtle differences: The American-made helmet uses a thicker wire for the chinstrap loops and a round-head rivet to secure the loop to the helmet shell. The British helmet relies on a split rivet and thinner wire.

registration numbers. (This is not always a fool-proof method. The author possess a British-made steel shell bearing the initials "HS 40" with an American manufactured and marked liner!)

- The texture on many British Mark I helmets is a mixture of paint and sand, whereas American M-1917's are textured with sawdust. (This becomes complicated when later versions of the Mark I were also textured with sawdust.)
- The chinstrap loops on the American M-1917 are more robust than the British Mark I.

Occasionally soldiers would camouflage their helmets. This was not an official practice, nor was it as widespread as many believe. Many of these "personalized" helmets were customized as soldiers returned to the United States on troop ships. Others were adorned with unit insignias and names of battles as remembrances after the conflict had ended. (The Leffel Collection)

It was very common for American soldiers to write their name on their helmet. Jake Abram, assigned to Company I, 342nd Infantry didn't want anyone to walk off with his "steel pot." Collectors should also closely examine chinstraps as many "doughboys" stamped names and units into the leather. (The Leffel Collection)

Corporal Cecil R. Laws, 30th Company, 5th Regiment, U.S. Marine Corps, with a regulation pack. The photos were taken in Paris on April 20, 1918. During this time frame it was not common to carry the eagle, globe and anchor emblem on the helmet.

Occasionally, the USMC Eagle/Globe/Anchor emblem, as used on the service cap, was affixed to the helmet. In some cases a unit insignia was painted below the E/G/A emblem. This painted insignia was also used without the metal emblem. Left: E/G/A emblem affixed to the helmet. Above: A painted unit insignia of the 5th Marine Brigade positioned below the E/G/A emblem. (Courtesy of the Col. Barry Smith collection)

This soldier is equipped with experimental body armor. The chainmail facemask is attached to the small hooks welded to the top flairs of the helmet. The lengths of chain seen hanging at the sides were for adjusting the fit of the mask on the helmet. (National Archives photo)

This front and rear view display a soldier equipped with an experimental armored neck and face protector. The rear view provides a clearer look at the hooks used to secure a chainmail facemask onto the M-1917.

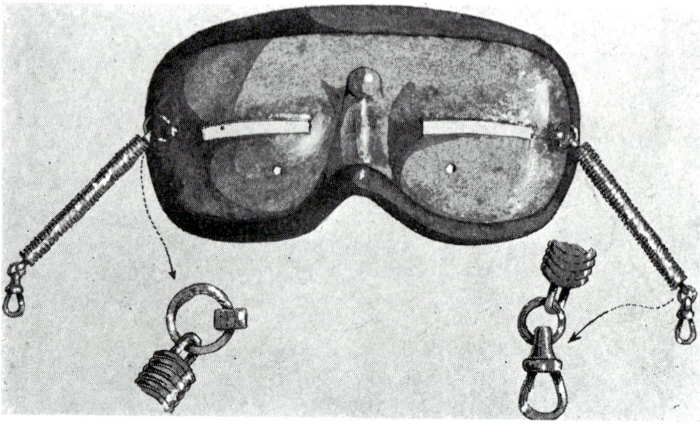

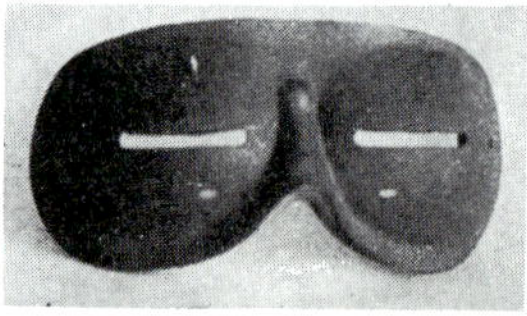

A Wilmer-model eyeshield which could be attached to the M-1917 helmet.

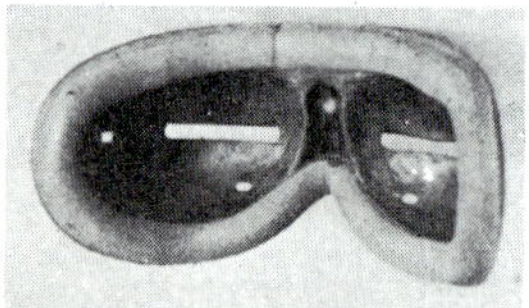

An American-made M-1917 helmet with dates, locations of battles, and sayings of the day. On each side of the helmet, three small holes have been punched or drilled. These holes were used to attach the Wilmer eyeshield attachment to the helmet.

A closer look at the holes positioned at the side of the M-1917 helmet. The single wire loop is probably the remnant of a Wilmer eyeshield. The solid head rivet and heavy chinstrap loop confirm this is an American-made M-1917 helmet. (The R. James Bender collection)

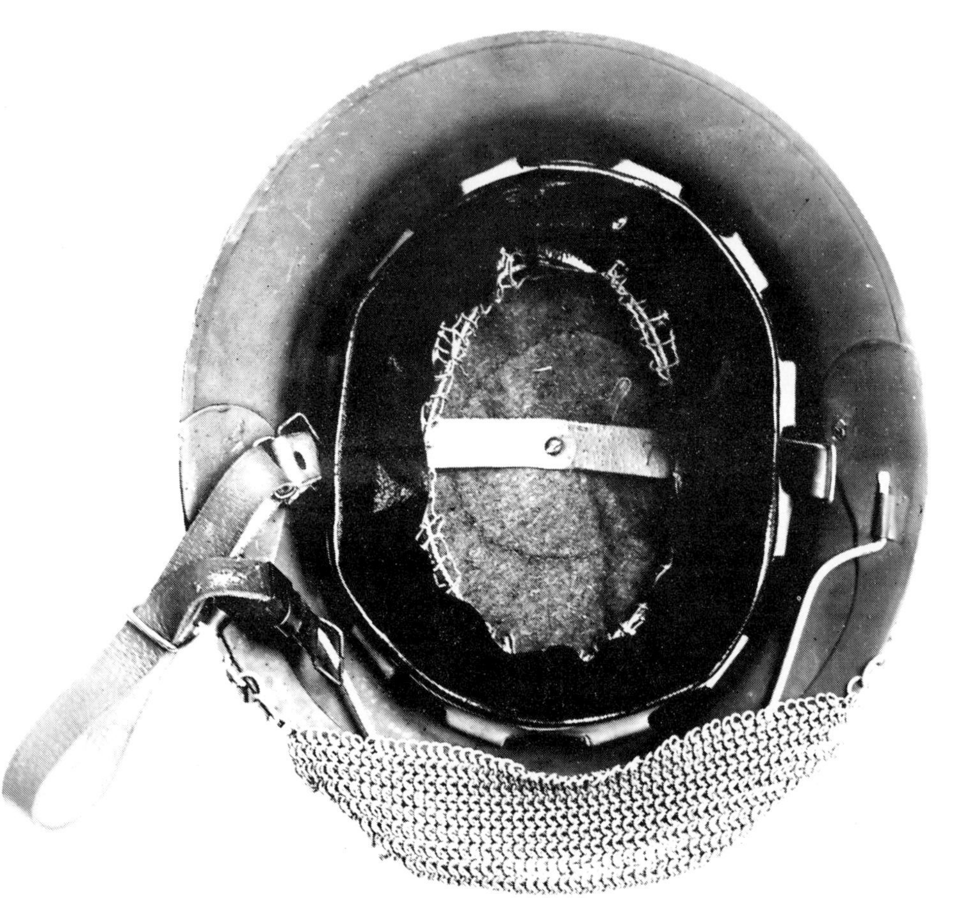

This photograph shows a M-1917 helmet with a chainmail facepiece in place. The thick wire was used to keep the mask away from the face. It is interesting to notice the side flairs of this helmet have been reinforced. The reverse of this helmet would show hooks welded on the outsite flair where the ends of the mask would be attached. (Photo Courtesy of Rock Island Arsenal Musuem)

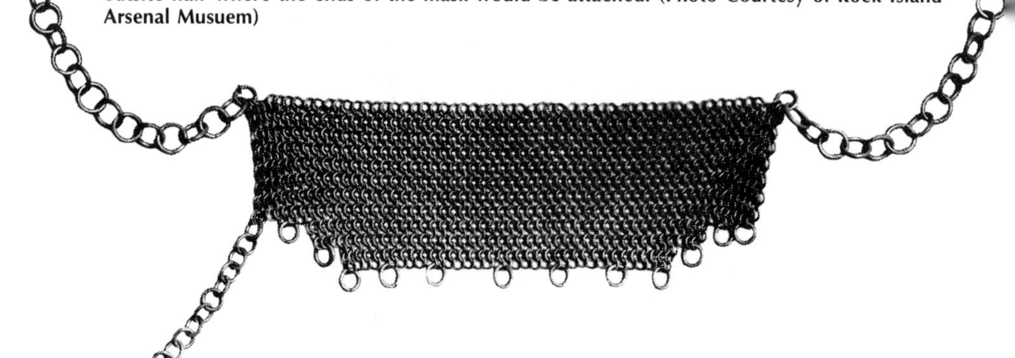

A chainmail face mask used on the M-1917 helmet. These facemasks, while not widely used saw action with some machine gun crews and at times on tanks to prevent "lead splash." Lead splash refered to small particles of metal flying off the inside of the tank body as bullets and projectiles struck the outside of the tank. As you can see, the chain could either be attached to the helmet using the hook method or through holes in the helmet shell. (Photo Courtesy of Rock Island Arsenal Musuem)

This altered M-1917 was obtained directly from a descendant of a World War I Division member. The reason for conversion can only be speculated. Perhaps it was decided that as the first American black soldiers landed in France they should already be outfitted in French-looking uniforms made in the United States.

Left: Note the French-style helmet on a World War II 93rd Infantry Division shoulder patch.

Black Americans had to wage a war at home to fight the enemy overseas. Eventually an all-black combat division was formed, the 93rd Division. America's military leadership lacked confidence in the 93rd and equipped the unit with out-dated and secondhand equipment. A distinctive feature which set the 93rd Division apart from its fellow American units was the helmet worn by its soldiers. The French helmet and French style uniforms were used by many soldiers assigned to the Division throughout the war. In fact the helmet became a source of pride for these American soldiers. The French helmet became the symbol for the Division and its silhouette adorned the official insignia for the Division. The helmets shown in these two photos are interesting in that they are actually a conversion of a standard M-1917 crown with new sides or flair attached. This alteration resulted in a helmet that looked similar to the French helmet, but was slightly shallower. (Courtesy of the Mac West Collection)

Note that the inside of the crown is still in its M-1917 brown color.

An interior view of this conversion helmet. It is easy to see the M-1917 helmet suspension and shell crown. The sides, or flair, of the helmet were added separately. No one knows how many of these conversion M-1917/French helmets exist. (Courtesy of the Mac West Collection)

An interior look at a standard French Adrian helmet. The suspension consisted of six leather leaves. The chinstrap was riveted in place and was also made of leather. Most members of the 93rd Division wore the standard French helmet.

This is an example of the standard issue blue painted French Adrian helmet from World War I. While it is a much better looking helmet than the M-1917 it failed to provide as much protection because the armor was much thinner. It should be noted that when worn by American soldiers, there usually was no insignia on the front of the helmet. (Courtesy of the Don Leffel Collection)

Proudly wearing their French-made helmets, members of the 369th Infantry Regiment of the 93rd Infantry Division prepare to display the colors. France, May 5, 1918. (Photo courtesy of the U.S. Army Military History Institute)

CHAPTER TWO

THE ART OF WAR: THE EXPERIMENTAL U.S. HELMETS 1917 TO 1936

The decision to adopt the British-style helmet did not stop efforts to secure a distinctive American-made helmet. The deficiencies of the M-1917, coupled with fears of insulting the French by not adopting their design, made helmet development an important mission of the Ordnance Department.

From the earliest days of America's entry into the First World War, the mission for developing experimental helmets belonged to Major Bashford Dean and the Armor Committee of the American Council of National Research, which he chaired. Dean, who was given the rank of Major by the Army and assigned to the Ordnance Department, was curator of the Department of Arms and Armor in the Metropolitan Museum of Art in New York.

Bashford Dean was curator of Armor at the Metropolitan Museum and was appointed to the rank of Major in the US Army Ordnance Department. He had a passion for armor and worked diligently experimenting with a variety of designs in an effort to develop a distinctive American combat helmet. In the end, Dean's primary contribution to US helmet development was that his designs clearly showed the superiority of a pot-shaped design. His research, and the prototypes which he and his staff at the Metropolitan Museum created, were studied extensively in the 1920s and 30s and helped propel the Infantry Board towards adoption of the Pot-shaped M1 helmet. Dean died in 1928 at the age of 61. He is remembered as an innovator in experimental helmet design and his book, Helmets and Body Armor in Modern Warfare, has been the premier reference for information on helmets of the early 20th Century. (Photo courtesy of the American Museum of Natural History)

Dean's position at the museum and knowledge of historical armor made him an ideal candidate to manage the development of prototype armor. The Metropolitan Museum graciously placed its Arms and Armor Department and its attendant craftsmen at the disposal of the government to support the war effort. Joining Dean on the Armor Committee were scientists skilled in the composition and development of metals, fellow students of armor, and industrialists.

Initially in the pursuit of a suitable American helmet, the Armor Committee looked overseas and inspected French and Belgian helmets. It also appears the Committee examined the exceptional Model 1916 helmet worn by the German army. In the end, it was decided that if a helmet other than the M-1917 was to be adopted, it would be a home grown American creation.

From 1917 until shortly after the First World War, the Ordnance Department, with the artists of the Metropolitan Museum, created no fewer than 16 helmets for consideration by the Armor Committee. Several of these designs were given serious consideration. No fewer than three models were tested extensively with between 1,000 and 1,800 examples of each produced. Several designs found their way to Europe for testing on the battlefield. Of the experimental prototypes, many were created only in a non-ballistic material.[1]

Even a cursory glance at the helmets which grace the following pages is proof that continuous improvement of helmet design was a very serious undertaking. Many of the helmets bear a close resemblance to armor worn during the 15th and 16th century. Deep profiles, round contours, visors and view slits were tried and tested. Many of the helmet designs appeared to hold great promise. The Model 2, Model 5 and Model 10, although rejected, appeared overall to be solid designs.

Some designs, like the Model 8 with its deep visor, appeared radical but were actually very well-balanced and comfortable.

On the other hand, some examples were destined to fail, such as the enormously heavy Model 7.

American soldiers and the opinions they had of the experimental helmets also had a great deal to do with acceptance or rejection. Almost without question, any helmet which bore the slightest resemblance to the German helmets was dismissed. The Liberty Bell design was very nearly approved as the American-made replacement of the M-1917. The soldiers however, disliked the appearance and that helmet, too, went by the wayside.

Many of the helmets featured on the following pages are believed to be one-of-a-kind examples. Some, like the Model 2 and the Model 8 were produced in sufficient numbers that from time to time an example is sold at a collectors' show or military auction. Others, like the steel tankers helmets of the 1930's, remain a mystery. Almost nothing, but for the photos, is known about these articles.

The attentive reader will no doubt be struck by the similarity so many of these helmets have with our current Kevlar-constructed ballistic helmet.

---

[1] The non-ballistic material was very thin sheets of steel used to create a prototype helmet. While it looked like an actual helmet, the steel used would not resist the impact of a bullet.

Many attributes of the experimental helmets of the post-World War I era continued to evolve and adapt, albeit in a material stronger than steel.

Many of the helmets pictured on the following pages rest in a dark, secluded storage room, in a World War II-era building at the Ordnance Museum of Aberdeen Proving Ground, Maryland. According to Ordnance Museum records most, if not all, of the examples stored at the Ordnance Museum, were provided by Major Bashford Dean.

While Dean made many contributions to helmet development, it must be remembered that he was first and foremost involved in experimental design. While it was Dean's ambition to create the American combat helmet, in the end, none of his creations, or those developed by the Ordnance Department and the Metropolitain Museum, ever found their way into active service with the U.S. military.

Helmet development continued on a sporadic basis after World War I. In the 1920's several modifications and improvements were made to the Experimental Model 5A for infantry use. These came to nothing as the Army chose to stay with a modified design of the M-1917. Other projects included obscure attempts to create a steel protective helmet for tank crew members. This too eventually fell by the wayside.

An interesting design considered by the U.S. Army is this standard French helmet with a "Polack" visor. This visor dates to approximately 1917. Several variations of Polack visors were made. This photo shows the visor in the down position. (U.S. Army Ordnance Museum, photo by author)

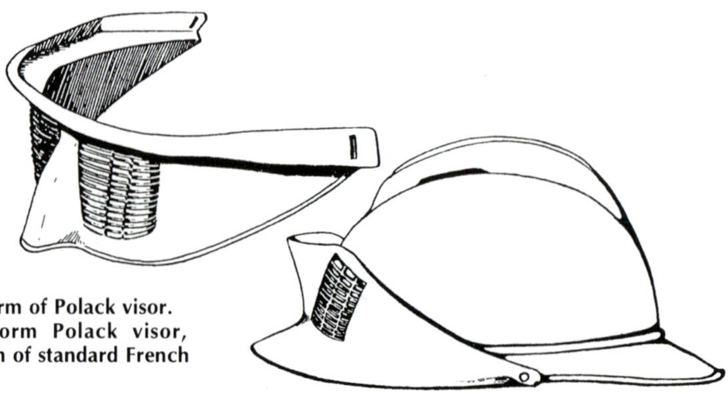

Above: Early form of Polack visor. Right: Early form Polack visor, adopted to brim of standard French helmet.

The Polack visor-equipped French helmet was tested by the United States and Britain. While the U.S. opted not to adopt this helmet, the French government did provisionally accept it and an unknown amount were produced and used in combat. (US Army Ordnance Museum, photo by author)

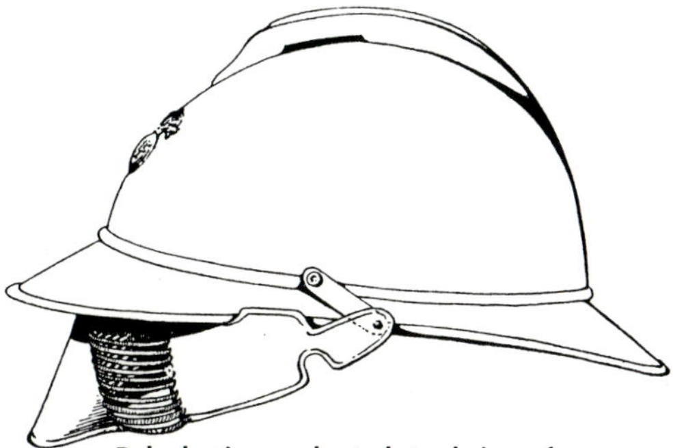

Polack visor adapted to brim of standard helmet.

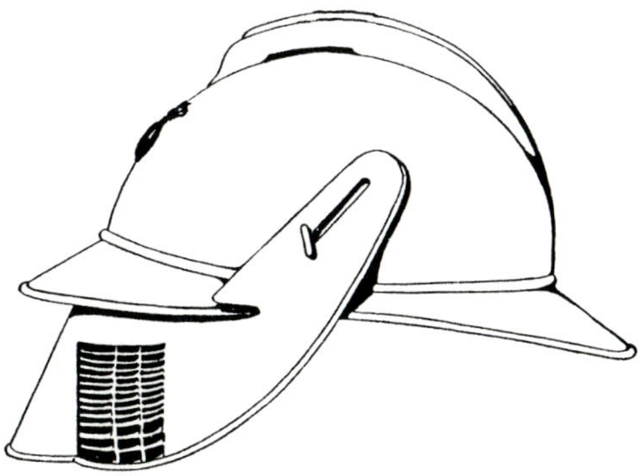

Polack visor, fitting above or below brim of helmet.

This photo, taken at Fort Polk, South Carolina in May 1918 shows a U.S. soldier wearing the U.S. copy of a French style helmet equipped with a Polack Visor. France, England and the U.S. were evaluating this style of helmet.

An interesting but undesignated helmet which could function as both armor and an entrenching tool. The unidentified individual in the photograph is believed to be the actual designer for this unique helmet.

This Franco-American designed helmet was also considered for adoption. Like other helmets of the day, it incorporated a Dunand visor system. (Photo courtesy of David B. Powers)

The liner and chin-strap system of the Franco-American helmet project. The chin-strap was entirely leather. It appears the strap was threaded through a loop and riveted in place. The suspension is a leather hammock style and relies on nine leaves to provide support. The drawstring, used to secure the leaves together, is missing in this example. It is unknown how many of these helmets were produced. (Photo courtesy of David D. Powers)

A rear view of the Franco-American Dunand helmet. Although smaller in size, the helmet did appear to provide more protection than the M-1917 helmet. (Photo courtesy of David B. Powers)

A front view of the Franco-American Dunand visored helmet. The craftsmanship on this example is exquisite with a large U.S. national emblem hammered into the non-ballistic shell. (Photo courtesy of David B. Powers)

The US Army also considered the standard Belgian helmet as a possible candidate. Equipped with a visor, the Belgian helmet was very similar to the British helmet but slightly deeper. The visor was the creation of M. Jean Dunand who offered this helmet design to the U.S. Army. A close look at the photos reveals the markings "U.S.A" on the forehead area of this helmet. It too was ultimately rejected for American service. (U.S. Army Ordnance Museum, photo by author)

With classic helmet designs of 15th Century Greece in mind, this helmet, designated the Model 2, was created in June 1917. It was made from a steel mixture of 12% manganese. With the liner, the helmet weighed in at 43 1/2 ounces. Its dimensions were 8 1/2 inches deep, 10 inches wide, and 12 3/4 inches long. (U.S. Army Ordnance Museum)

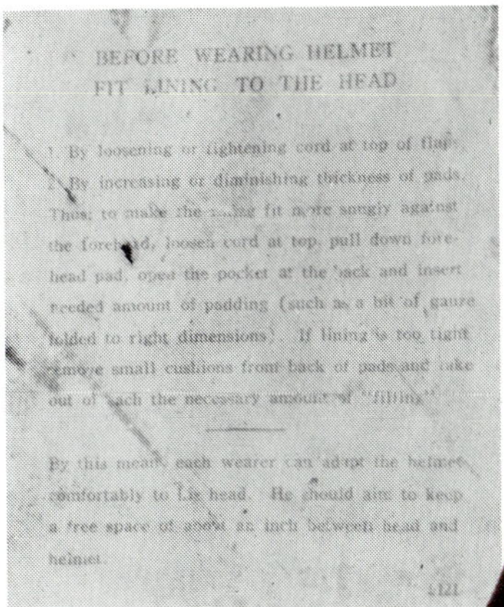

BEFORE WEARING HELMET
FIT LINING TO THE HEAD

1. By loosening or tightening cord at top of flaps.
2. By increasing or diminishing thickness of pads.
Thus; to make the lining fit more snugly against the forehead, loosen cord at top, pull down forehead pad, open the pocket at the back and insert needed amount of padding (such as a bit of gauze folded to right dimensions). If lining is too tight remove small cushions from back of pads and take out of each the necessary amount of "filling."

By this means each wearer can adapt the helmet comfortably to his head. He should aim to keep a free space of about an inch between head and helmet.

4121

An example of the tag glued to the inside crown of the Experimental Model 2. (Courtesy of Bill Rizzutti)

Approximately 2,000 of these helmets were made by Ford and Company in Philadelphia in the fall of 1918. This photo clearly shows the depth of the helmet. The front contours worked well to deflect projectiles. While the helmet was heavier than the M-1917, reports indicated it was far more comfortable to wear. (U.S. Army Ordnance Museum)

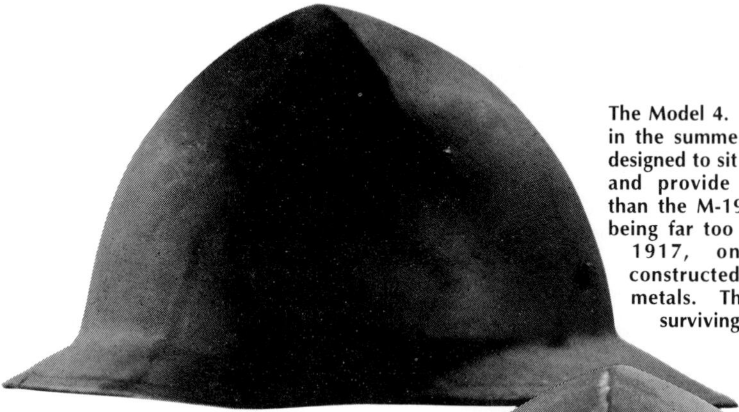

The Model 4. Initially conceived in the summer of 1917, it was designed to sit lower on the head and provide more protection than the M-1917. Criticized as being far too similar to the M-1917, only a few were constructed out of lightweight metals. This may be the only surviving example.

Another view of the Model 4. This helmet was designed at the Metropolitan Museum by Daniel Tachaux. The purpose of the lugs attached to the steel shell is not clear. They may have been used with a visor system. The helmet used a French style of suspension. (U.S. Army Ordnance Museum)

The three-pad suspension system of the Model 5 style helmet. Like the German M-1916, it was secured together in the center by a thong. Notice the experimental webbed chinstrap and its new buckle system. (Photo courtesy of David B. Powers)

This helmet is the Model 5A. This helmet was greatly influenced by the German Model 1916 design. Weighing in at 2 pounds, 12 ounces, the Model 5A was deeper and provided much better protection than the M-1917. The Model 5A was made of Hadfield Manganese by Hale and Kilburn of Philadelphia.

The Model 5 held great promise as a serious contender for the U.S. helmet. Unfortunately, its similarity to the German helmet made it an unpopular choice with the Army's command elements. (U.S. Army Ordnance Museum)

Different stages in the "drawing" or pressing of the steel used to form the Experimental Model 5 series of helmets. These helmets were manufactured by Hale and Kilburn in Philadelphia, Pennsylvania.

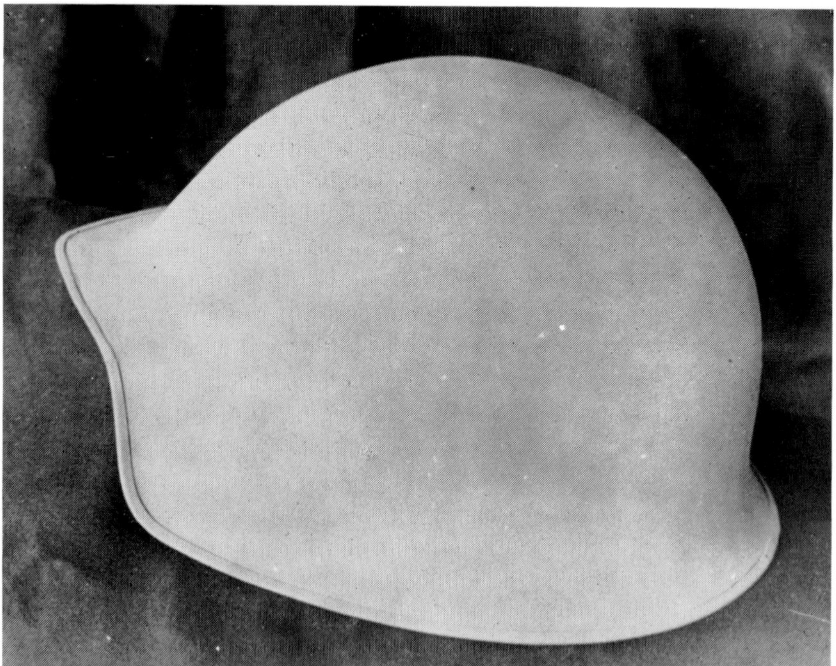

The finished product: The Experimental Model 5 hot off the presses at Hale and Kilburn, circa 1918.

The Experimental Model 5A was cousin to the Model 5. The primary differences were the gentler slope to the side and slightly deeper bowl. Several versions of the Model 5A were manufactured. (Photo courtesy of David B. Powers)

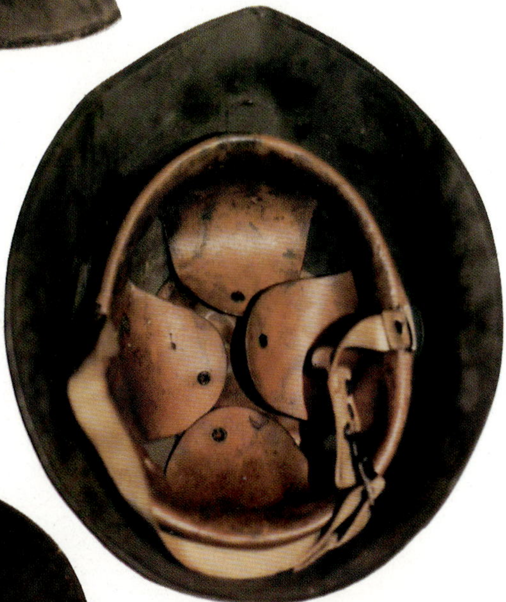

A look at a modified suspension system designed in the 1930's for the Model 5A. The system uses four leather leaves which would be tied together at the crown. Beneath the leaves, a round leather pad provided a cushion. This type of suspension would eventually be used in the M-1917A1. (Photo courtesy of David B. Powers)

Another version of the Model 5A. A close comparison between this example and the previous Model 5A shows the rear of this helmet is "cut-up" and does not extend as low. (Photo courtesy of David B. Powers)

Another interesting example of the Model 5A. Notice that on the rim from the chinstrap back, the rim of the helmet is curled up, somewhat like a rain gutter on a house. The specific purpose of this is not known. In the mid 1930's, the U.S. Army Infantry Board at Fort Benning tested and compared the Experimental Model 5A variants against the M-1917 and M-1917A1. (Photo courtesy of David B. Powers)

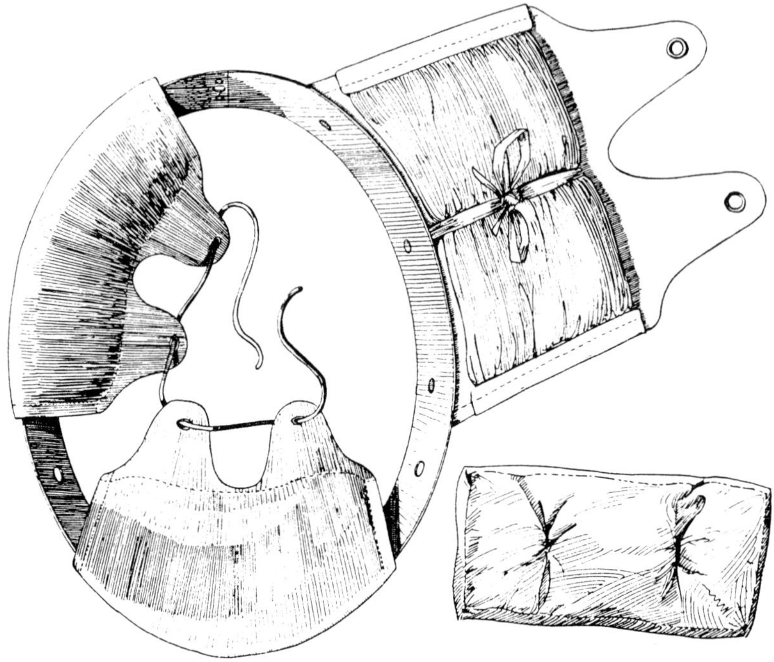

A diagram of the Model 5 and 5A's suspension system. The three leather pads were attached to a steel band which ran around the interior of the shell.

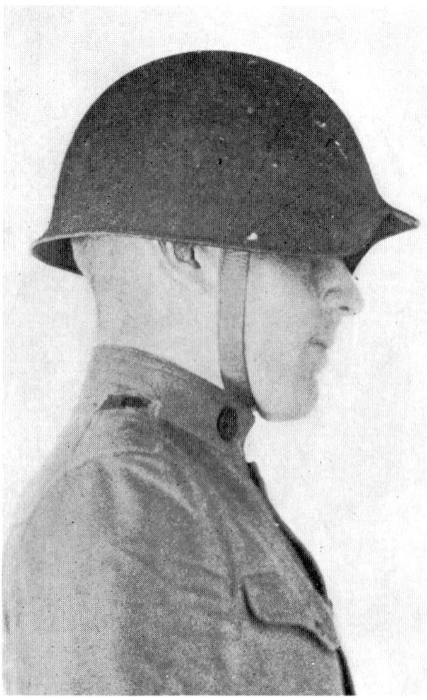

American experimental helmet No. 5.

Developed by Daniel Tachaux, the Model 6 was produced purely as an experimental helmet. With its interesting design, the helmet could be worn in the manner shown in the photo or with the face plate tilted down. Obviously, with the faceplate lowered vision would be seriously obscured. This example was made in a non-ballistic metal.

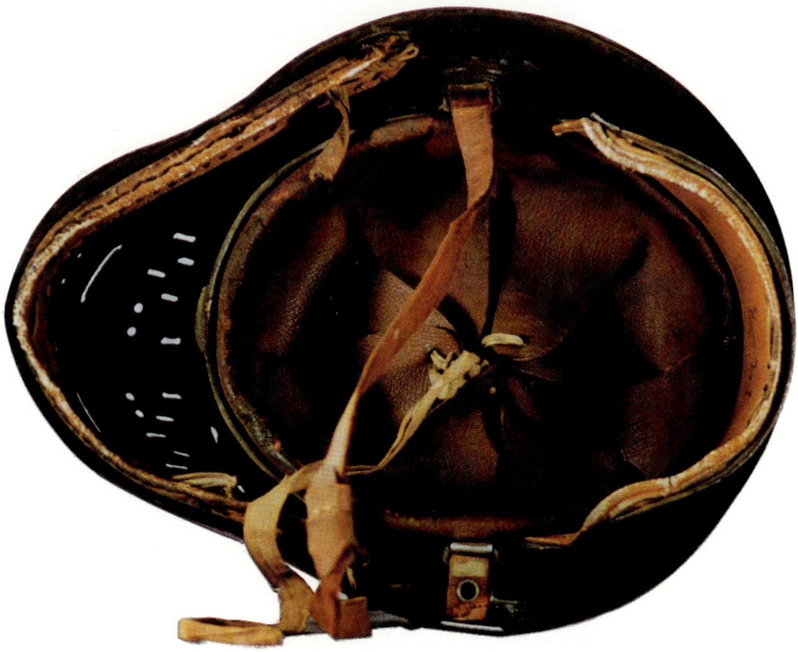

An interior view featuring the suspension system of the Model 6 helmet. The Model 6 was very different from other experimental helmets produced at the time. To include the faceplate, the interior was covered in padded leather. The Model 6 is probably a one-of-a-kind helmet. One can only imagine it would be very heavy and cumbersome to actually wear in combat. (U.S. Army Ordnance Museum)

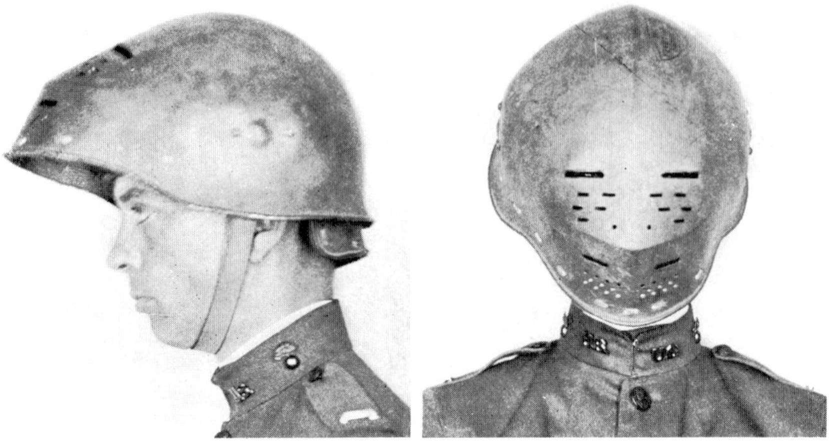

Experimental helmet (No. 6) with tilting dome.

Modeled after siege helmets of the 17th Century, the Model 7 was specifically designed to protect machine gunners and forward observers. Less than 50 of these helmets were made. They were, however, made in three different weights: 11, 15, or 18 pounds. The helmet was actually tested in France but results were less than flattering. The weight of the helmet made it impractical to use in the field. These helmets were made by the William H. Mullins Company of Salem, Massachusetts. (U.S. Army Ordnance Museum)

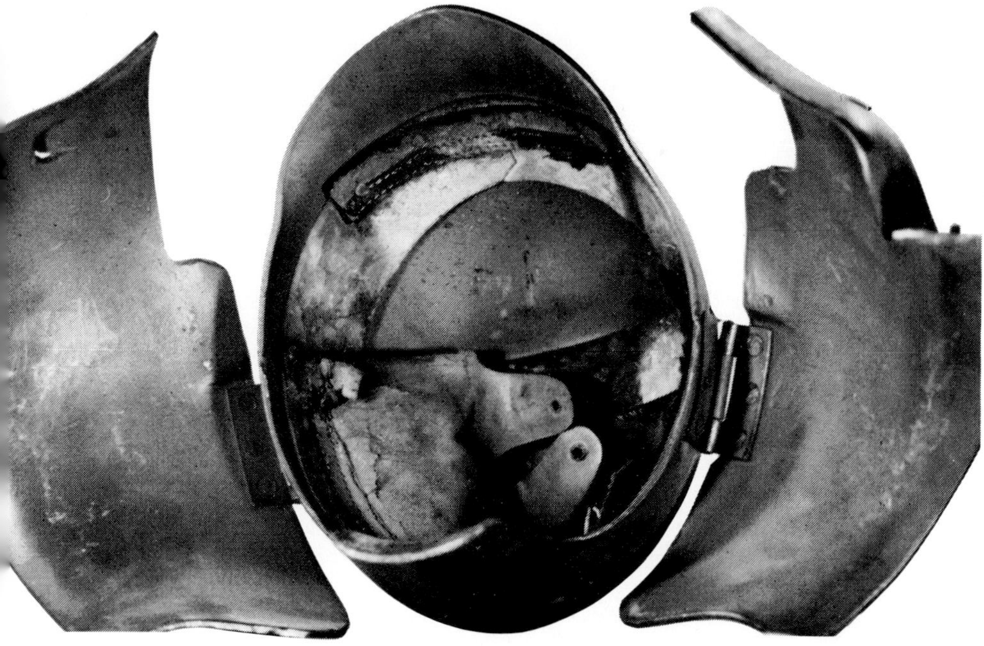

A look at the three pieces of steel which made up the Model 7. A skullcap with a long tail flared from the rear of the helmet to protect the back of the head and neck. The other two parts were the "clamshell" face plates which were riveted to the shell on hinges. The plates hung from the sides of the skullcap. The interior of the Model 7 relied on a suspension similar to the German M-1916 helmet. The protrusion near the occupants nose was an attempt to minimize condensation caused by breathing.

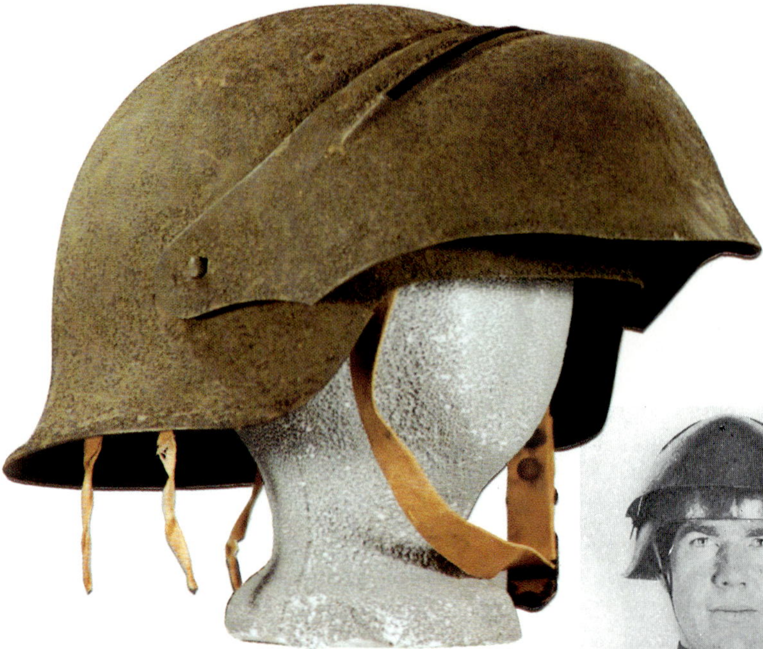

The fierce appearance of the Model 8 with its steel visor makes this a very distinctive and attractive helmet. Ford made 1,300 in November, 1918. Like many U.S. helmets, the Model 8 used a suspension system similar to the German M-1916. The helmets were constructed of Hadfield Manganese and weighed in at just over three pounds. The chinstrap was a new design constructed of webbing. It would later be used on the M-1917A1 and the M1 helmet.

The Model 8 could be worn with the visor in the up or down position. The helmet was painted a dark olive drab and was textured with fine particles of sawdust like the M-1917. The Model 8 is a sturdy, well built helmet. Unfortunately it received little interest and was eventually discarded. (U.S. Army Ordnance Museum)

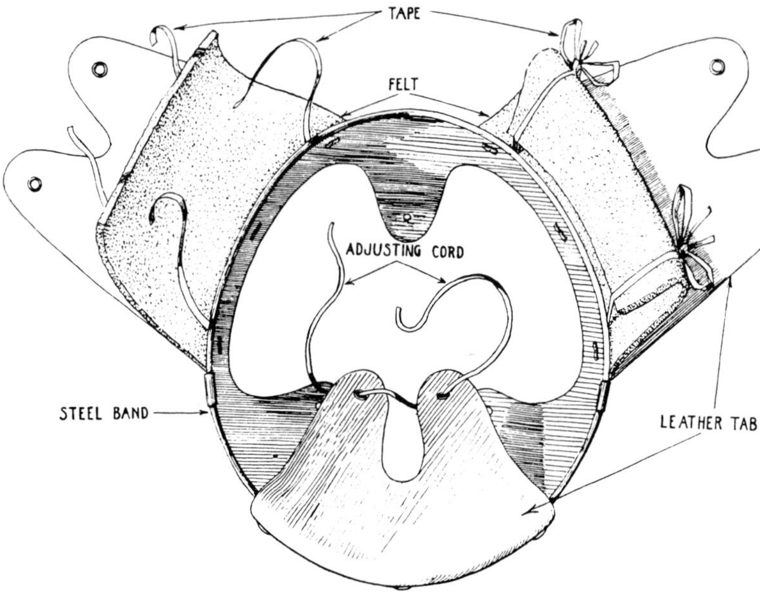

Fig. 171. Carrier of helmet model No. 8, showing lining pads or tabs

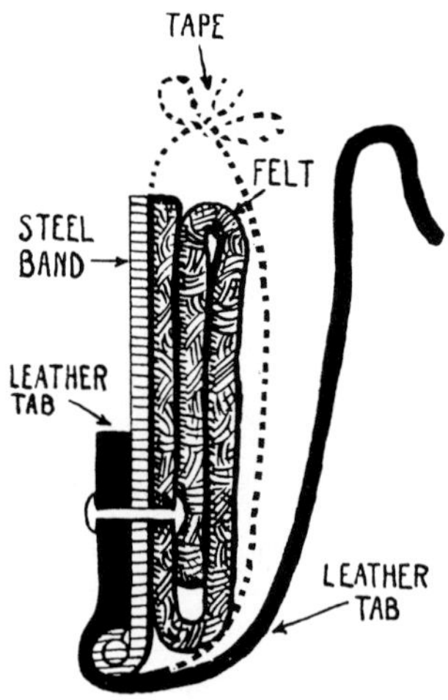

Fig. 172

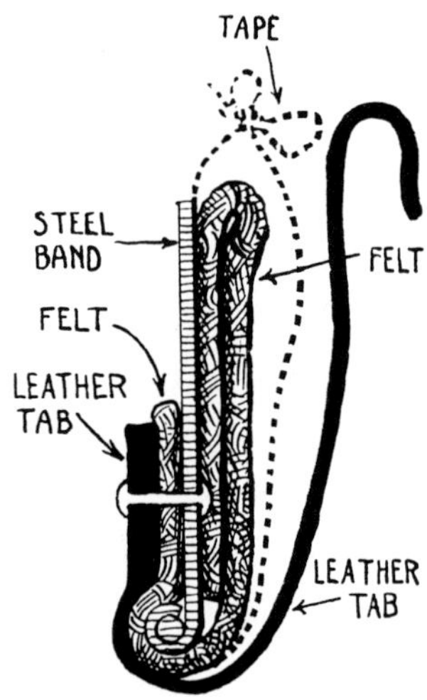

Fig. 172A

Fig. 172. Section of lining carrier showing arrangement of tabs for head sizes 7 and under, for 7⅛ and over

The internal suspension system used in the Experimental Model 8 helmet.

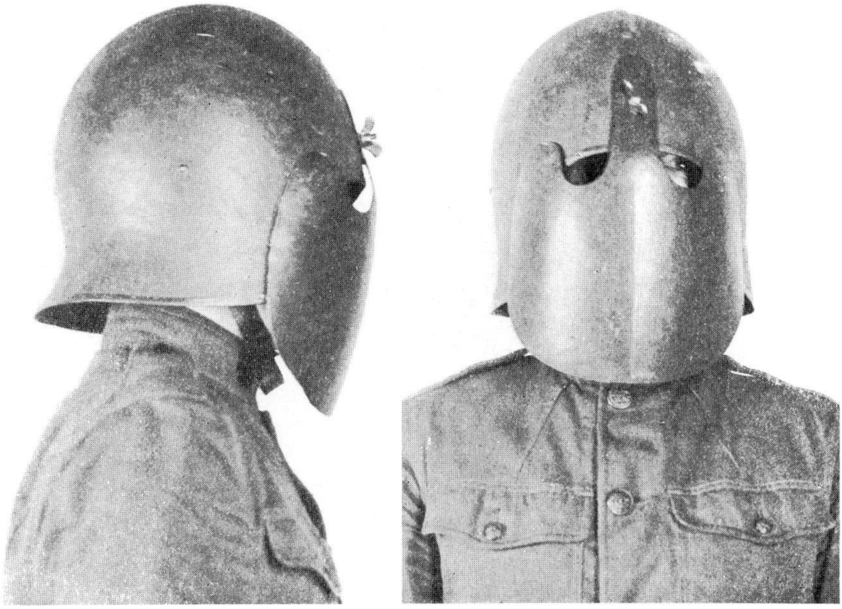

The Experimental Model 9. Very little is known about this helmet although it is obviously related to the Model 7. It is unknown if there any surviving examples.

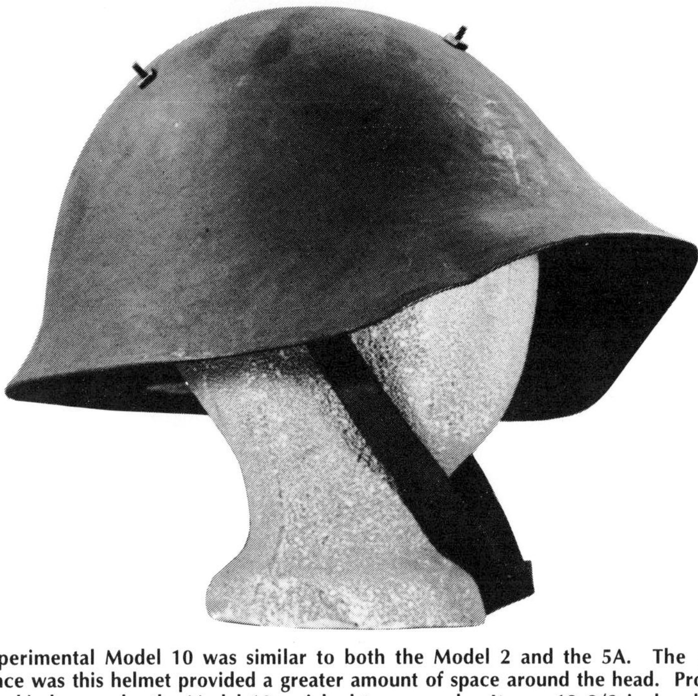

The Experimental Model 10 was similar to both the Model 2 and the 5A. The primary difference was this helmet provided a greater amount of space around the head. Probably a one-of-a-kind example, the Model 10 weighed two pounds. It was 12 3/8 inches long, 11 inches wide, and 6 3/4 inches deep. This hand-made model was constructed at the Arms and Armor shop at the Metropolitan Museum in 1918. (U.S. Army Ordnance Museum)

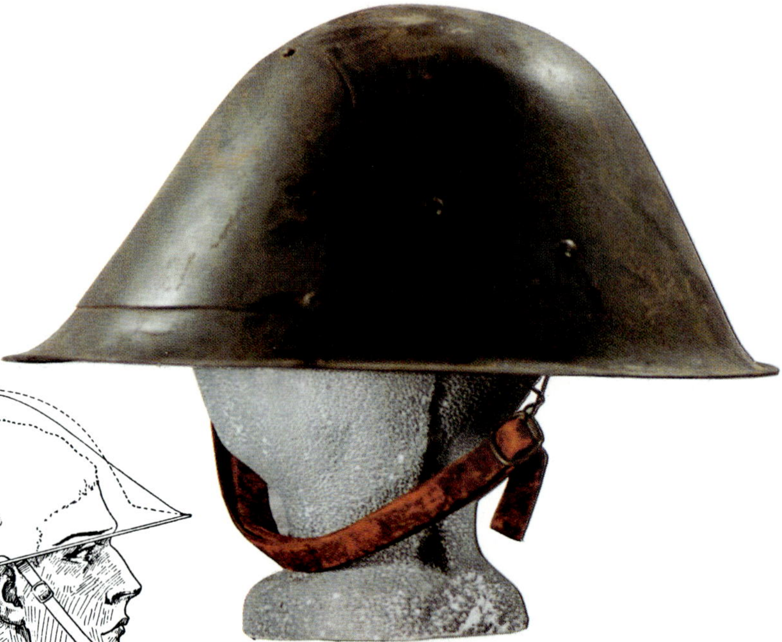

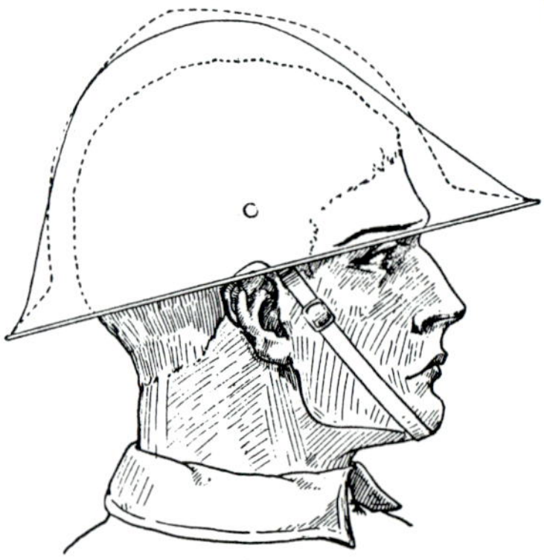

Fig. 181. Liberty Bell helmet. Fall, 1918. Shown over profile (in dotted line) of American experimental helmet model No. 4

This experimental helmet, designated "The Liberty Bell," and designed by Major James E. McNary, was submitted to the Helmet Committee for consideration and was initially accepted as the replacement helmet for the Model 1917. Essentially a variant of the Model 4, between 1,000 - 2,000 examples were manufactured in France at the request of the American General Staff. The standard Liberty Bell weighed 39 1/2 ounces. This example has a reinforcement plate riveted to the front of the helmet. The helmet had many weaknesses including a failure to fully protect the head. Several types of liner systems were used in this design. Its greatest downfall was, however, its unflattering appearance. In essence, American troops who encountered the helmet hated it! The provisional decision to accept "The Liberty Bell" was abandoned in 1920. (U.S. Army Ordnance Museum)

This Liberty Bell example does not have the additional armor plate on the front. This example made use of a liner similar to that used in the German M-1916 helmet. Notice the two rivets on the side. (Photo courtesy of David B. Powers)

The interior of the Liberty Bell helmet with the three-pad design suspension system. The chinstrap on this example is full leather construction similar to that used on the M-1917. (Photo courtesy of David B. Powers)

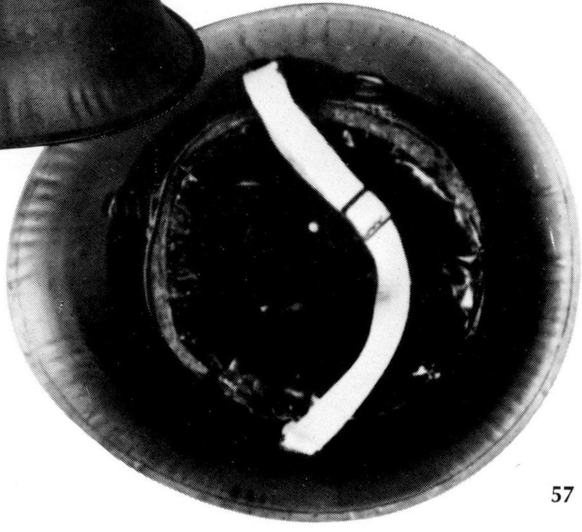

The Liberty Bell helmet uses a metal liner band, gray wool backing to the black oil cloth suspension. It has coiled springs as a buffer to each of the four sides. The lightweight tan cloth one-piece chinstrap uses a "H" buckle attachment. (Photo courtesy of David B. Powers)

According to Army Ordnance records, this helmet, designated the H-3-C, was designed by Bashford Dean, Chairman of the Helmet Committee and curator of the Arms and Armor Department of New York's Metropolitan Museum. The H-3-C was another attempt to incorporate a visor into the new helmet. This helmet is probably a one-of-a-kind example.

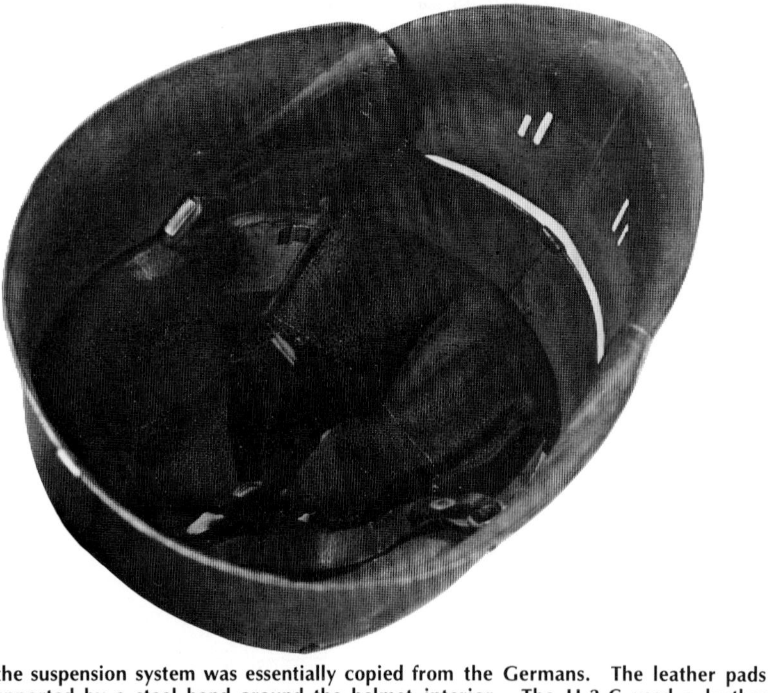

Again, the suspension system was essentially copied from the Germans. The leather pads were supported by a steel band around the helmet interior. The H-3-C used a leather chinstrap similar to that used on the M-1917. (U.S. Army Ordnance Museum)

In its visor-down position, the H-3-C would have caused serious vision problems for soldiers in the field. Consider how difficult it would have been to shoulder, aim, and fire a weapon with the visor down. Readers should remember the primary motivation for using visors was to provide increased protection to soldiers. Obviously, designers took this objective very seriously.

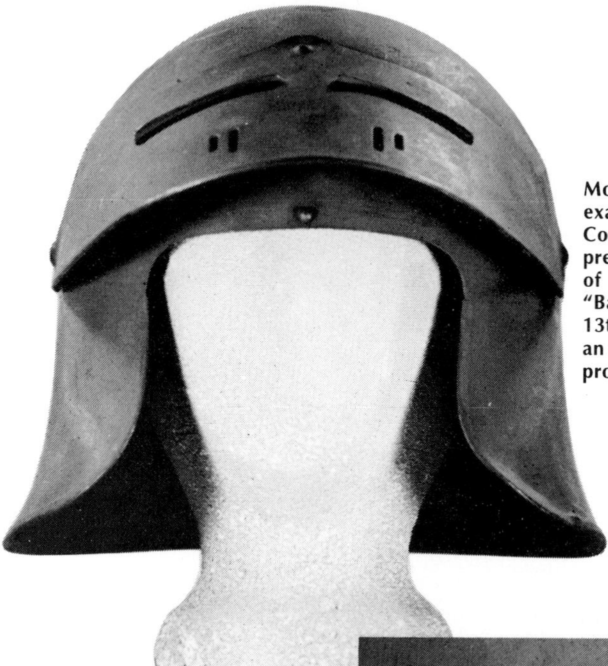

Most likely a one-of-a-kind example, this is the H-6-A. Constructed in a non-ballistic pressed steel. The basic shape of the helmet is similar to "Barbute" helmets from the 13th century. The visor was an attempt by designers to provide maximum protection.

The painted markings applied by the manufacturer of the helmet, Hale and Kilburn.

Although giving the appearance of providing all-around protection, the H-6-A had serious deficiencies. The shape of the helmet reduced both vision and hearing. Unfortunately, it appears there were no examples made in ballistic steel. Consequently, the concept of the H-6-A was never fully evaluated. (U.S. Army Ordnance Museum)

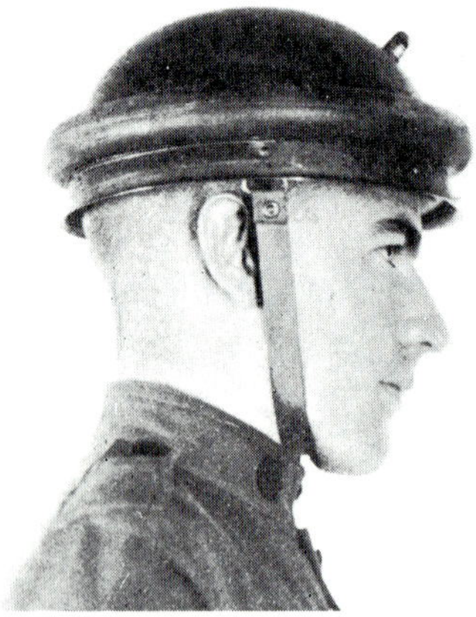

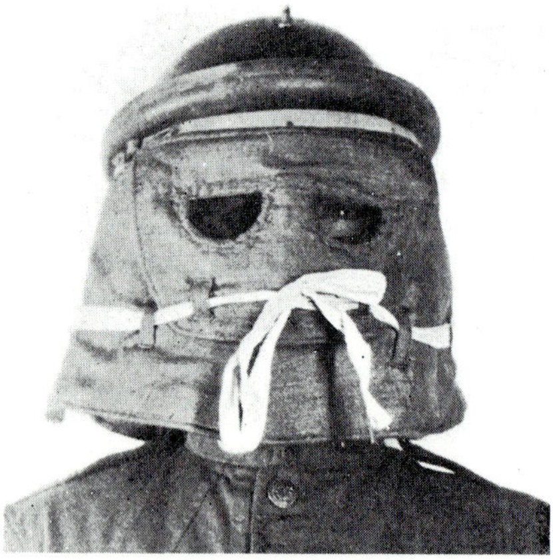

The Experimental Model 13 consisted of a steel skullcap wrapped in leather and padding. It was specifically designed for use by tank crew members. The helmet could be used with or without the special "splash guard." The guard protected the crew members from "lead-splash." Lead-splash was metal shards which broke loose from the inside skin of the tank as the exterior was struck by bullets and projectiles.

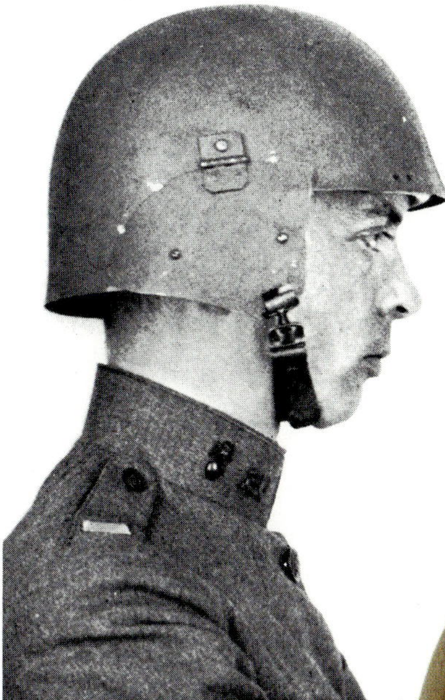

Efforts to improve the protection of soldiers extended into the air as well. This is the Experimental Model 14, a light steel helmet with ear flaps designed for aviators.

An improved version of the Model 14 was the Experimental Model 14A. Again, the helmet was constructed of very light steel and worn over the flight helmet.. Although flimsy in appearance, the Model 14A was rigid and could have provided favorable protection. It weighed approximately 1 1/2 pounds.

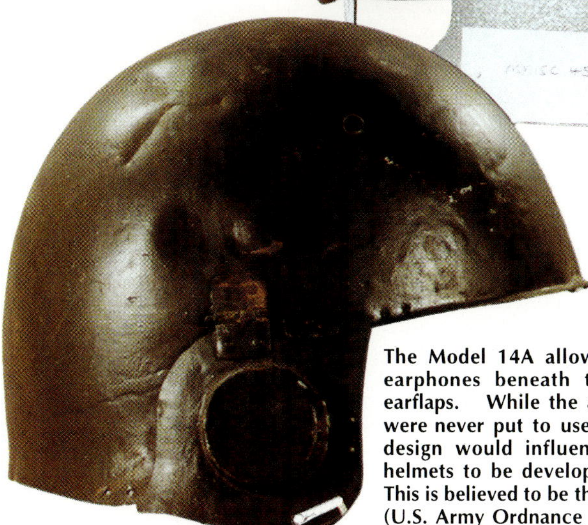

The Model 14A allowed the aviator to wear earphones beneath the cups formed in the earflaps. While the advantages of this helmet were never put to use during World War I, its design would influence a variety of anti-flak helmets to be developed during World War II. This is believed to be the only surviving example. (U.S. Army Ordnance Museum)

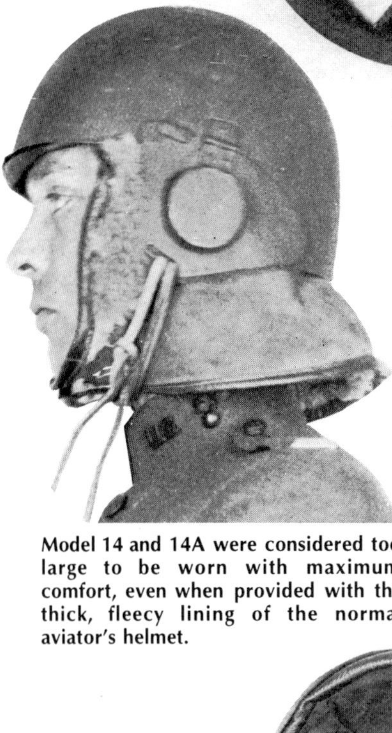

The Models 14 and 14A had a three-pad lining system.

Model 14 and 14A were considered too large to be worn with maximum comfort, even when provided with the thick, fleecy lining of the normal aviator's helmet.

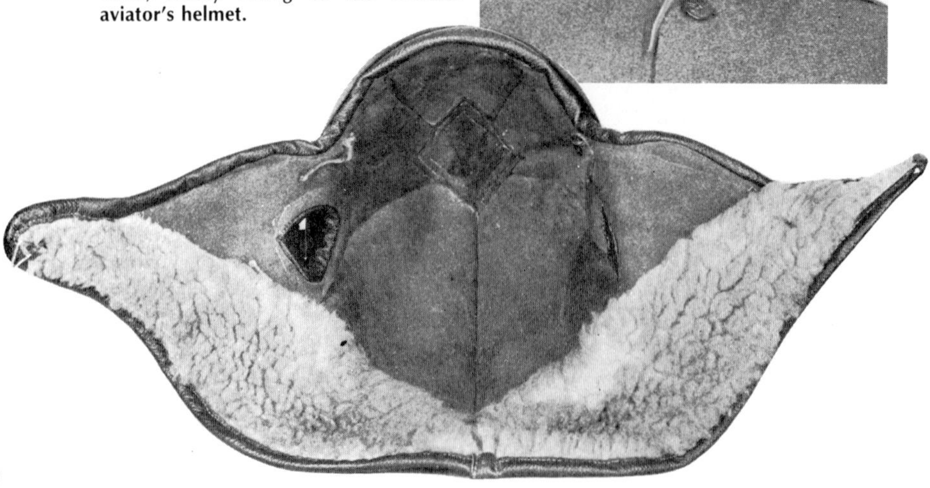

Aviator's experimental helmet Model 15, 1918. This was considered the best of the designs as it was closely modeled to the head with an intervening space of only one-half to three-quarters of an inch. It had a new-type chinstrap and a continuously cushioned type of liner.

A final burst of post-WWI helmet experimentation in the 1930s was due to the development and expansion of armored vehicles, and there was strong interest in the development of a specialized ballistic helmet for the crews. In 1934, the Infantry Board requested that the Ordnance Department develop a specialized steel helmet to replace the leather and fiber football-style helmets which had been in use. The criteria for the steel designs was quite strict; they had to provide ballistic protection equal to the M-1917, yet not interfere with the use of aiming devices. In addition, the new helmets would have to accommodate radio headphones plus goggles and gas masks. Although several test helmets were developed by the Ordnance Department, all were rejected by both the Infantry and Cavalry Boards. In 1936, the Ordnance Board closed the development and testing program stating that all the prototype steel helmets submitted for armored vehicle wear failed to meet the criteria. The program was dropped. Helmets, such as the experimental D777 example show the effort which was put into the project. While these interesting and obscure helmets failed to gain acceptance in their original form, several of the design concepts and elements would reappear in World War II anti-flak helmet designs.

During the post-war years, little if any activity in the area of helmet design or experimentation continued with the exception of comparisons of the Model 5A with the M-1917. However, it appears several designs for an improved protective steel helmet for tank crews were being considered. Identified as Experimental Model D7774, this helmet was designed in the mid-1930s. The author was unable to learn anything about this helmet. A keen eye will note the use of a heavy webbed chinstrap with the hook and arrow connector. Notice the chinstrap is riveted in place. The rings on the sides, it can be assumed, were to attach earphones to the helmet. (Courtesy of the Rock Island Arsenal Museum)

Unlike the previous helmet, this helmet was only identified as a Combination Tank Helmet, Experimental Model. Conceivably this may have been the "next generation" of the D7774 with its adjustable communications headset. The suspension installed in the helmet is obviously related to that used in the M-1917A1 "transitional" helmet. The suspension, like that in the M-1917A1, is secured in place using a screw-on nut rather than a rivet. (Courtesy of the Rock Island Arsenal Museum)

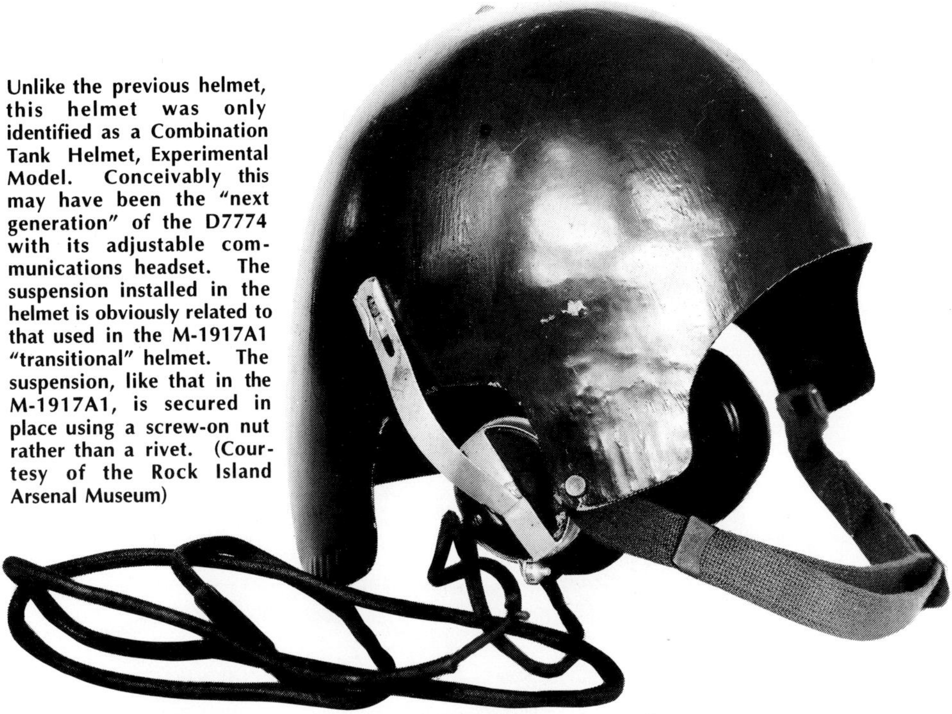

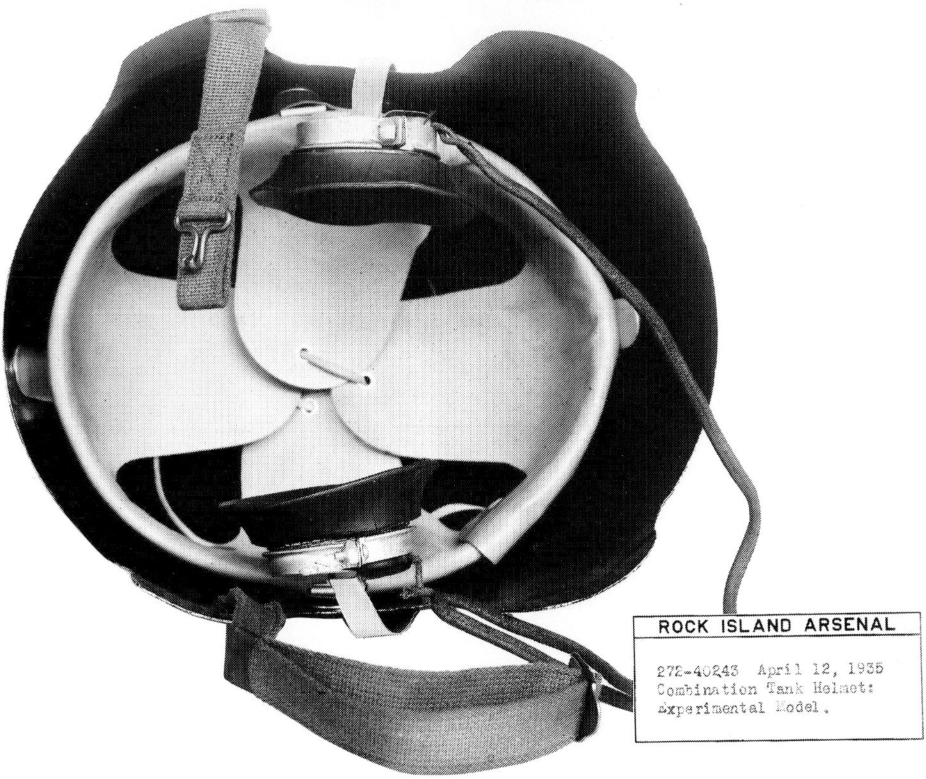

Another variation of the Combination Tank Helmet Experimental Model. This helmet provides both communications and protection by using an ear-flap system similar to that used on the Experimental Model 14 of World War I. (Courtesy of the Rock Island Arsenal Museum)

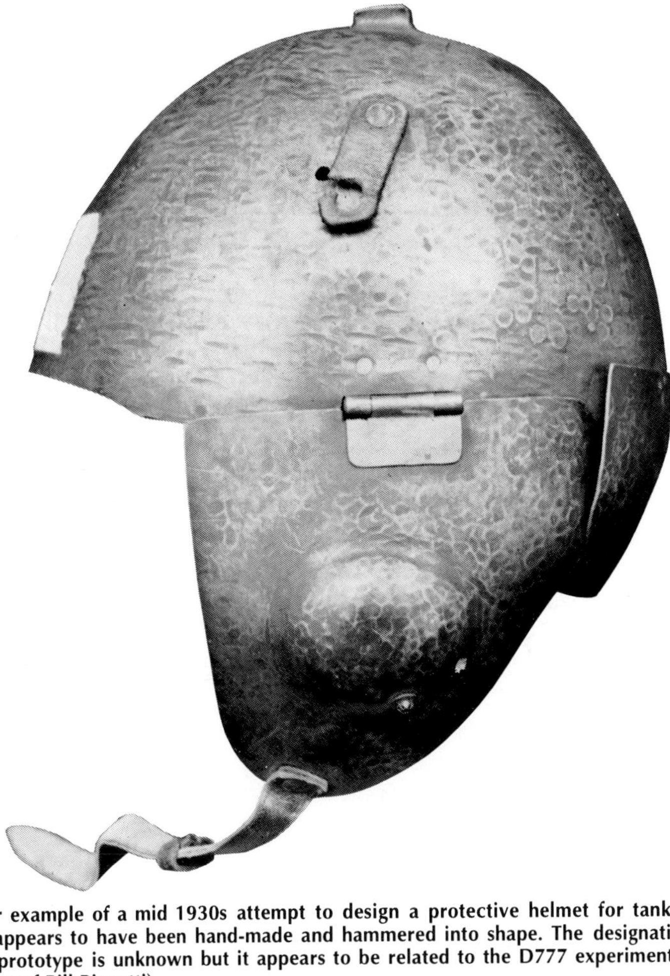

Another example of a mid 1930s attempt to design a protective helmet for tankers. This model appears to have been hand-made and hammered into shape. The designation of this helmet prototype is unknown but it appears to be related to the D777 experimental model. (Courtesy of Bill Rizzutti)

CHAPTER THREE

TRANSITION: THE MODEL 1917A1 HELMET

Immediately following World War I, the majority of helmet development work ground to a halt. The cessation of hostilities ended the interest in developing an American-made helmet. The work of the Ordnance Department, Bashford Dean and the Armor Committee of the American Council of National Research was essentially shelved.

The aftermath of World War I showed that the M-1917 helmet had deficiencies which needed to be addressed. The helmet offered limited protection to the side and rear of the head and was poorly balanced.

Recognizing the problems with the M-1917, the Infantry Board ordered an examination and evaluation of several of the Ordnance Department's experimental models in 1926. Determining that a deep, pot-shaped helmet could provide better protection for soldiers, the M-1917 was matched up against the experimental Model 5A at Fort Benning, Georgia.

Although it was deeper and covered more of the head than the M-1917, tests showed the Model 5A couldn't take the pounding of ammunition and shrapnel as well. In 1932, the decision was made to scrap further evaluations between the M-1917 and the Model 5A.

The decision to end further testing of prototypes did not change the fact the M-1917 had some serious flaws. With the prospect of a replacement seemingly gone, the Infantry Board decided to improve the M-1917.

Besides its limited protective ability, most of the criticisms of the M-1917 centered around its instability. Soldiers in the field constantly complained about the difficulty of balancing the helmet on their heads.

The Infantry Board responded by improving the suspension system of the helmet.

A Field Service Modification Order[1] was dispatched in 1936 ordering the alteration of all M-1917 helmets. M-1917 helmets would be fitted with the new liner and would be designated as Helmet, Model 1917A1.

It seems that rather than manufacture an entirely new steel helmet shell, existing M-1917 helmets were refurbished and the new liner installed. Along

[1] *Field Service Modification Work Order B-13-W1, January 1936*

The majority of M-1917A1 helmets were modifications of the M-1917. Consequently, many of the steel shells were refinished and repainted. This accounts for the variety of OD colors and textures which can be encountered.

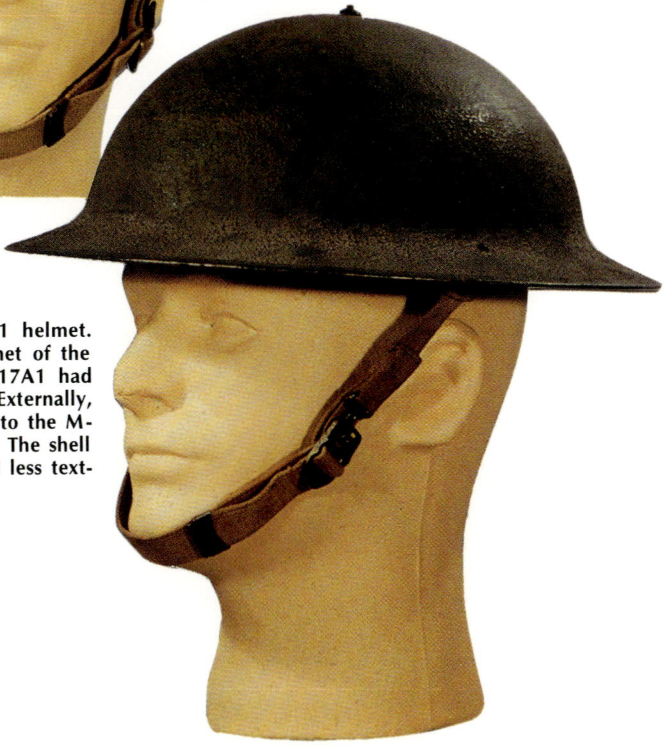

Side view of the M-1917A1 helmet. Although similar to the helmet of the First World War, the M-1917A1 had several distinctive features. Externally, the steel shell was identical to the M-1917 in shape and material. The shell was, however, smoother and less textured than the M-1917.

with the new liner system, the M-1917A1 would also be equipped with a canvas chinstrap.

Unlike the leather chinstrap with its sliding buckle mechanism, the M-1917A1 chinstrap would be secured with a hook-and-buckle device designed near the end of World War I.

The new liner system was composed of two materials: Steel and leather. Similar to the liner used in the M-1917, the M-1917A1 liner started with a thin steel band fitting around the head.

Two additional steel bands, bent to conform to the interior curvature of the steel helmet shell, one running from side to side, the other from front to back, would serve as the helmet's support system.

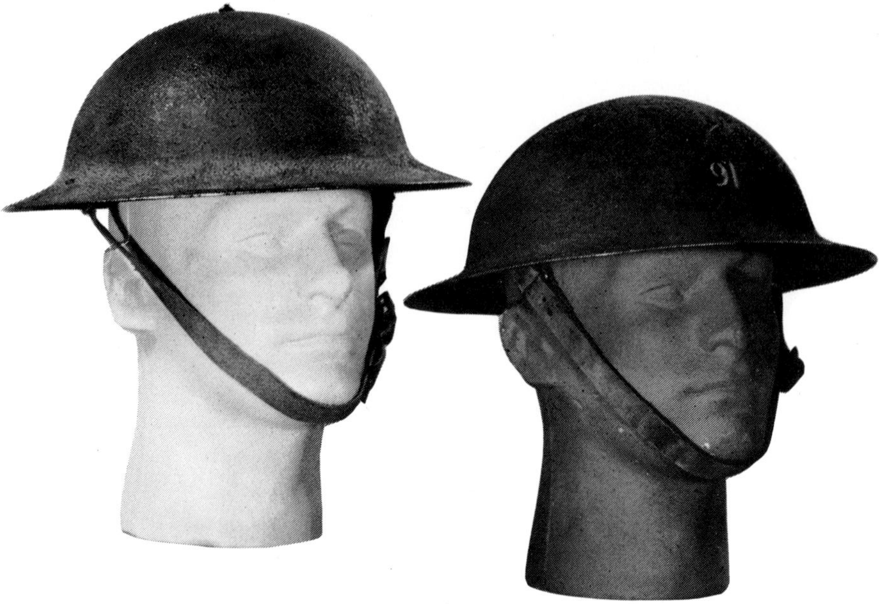

A comparison between the M-1917 (right) and the M-1917A1. The major external difference is the chinstrap. The M-1917A1 used a heavy stitched webbing, two-piece chinstrap. A nut visible on the crown, held the suspension system in place.

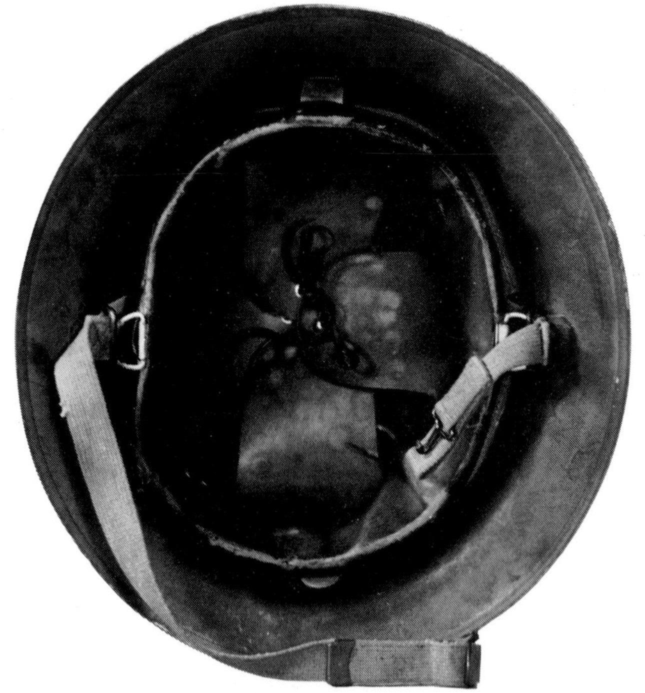

The new suspension system was at the heart of the conversion. The suspension used four metal strips bent to act as springs, providing tension to hold the new liner in place. Leather served as the main ingredient for comfort. In the crown, a small leather-covered oval pad was installed as a cushion. (The Leffel Collection)

Where the two steel bands met at the crown in the interior of the helmet, a single screw secured by a nut on the exterior crown of the helmet held them in place. This screw was one of the methods used to hold the liner's skeleton in place.

At the other end of the criss-crossing bands, three rivets secured them to the steel headband. Also taking advantage of these three rivets was another steel band, bent into a spring-like hook. These hooks, on the end of the criss-crossing bands, four in total, served as tension springs to help hold the liner in place.

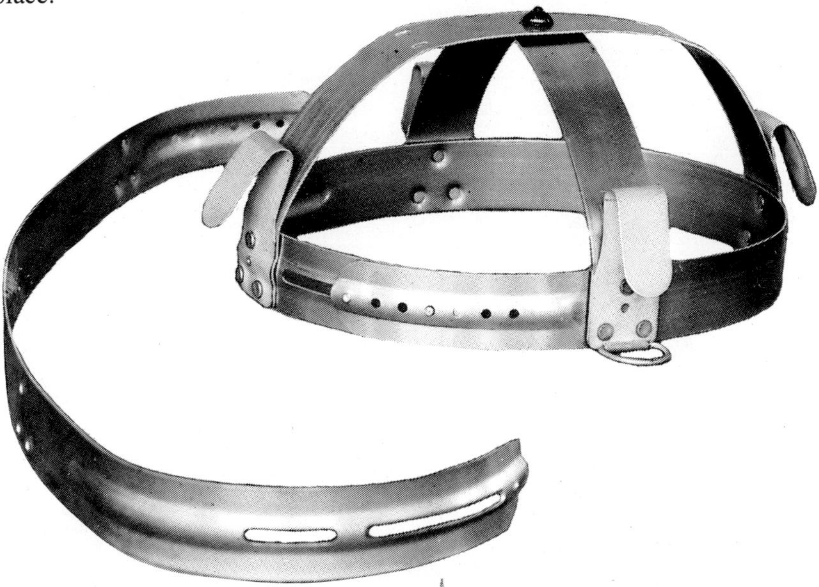

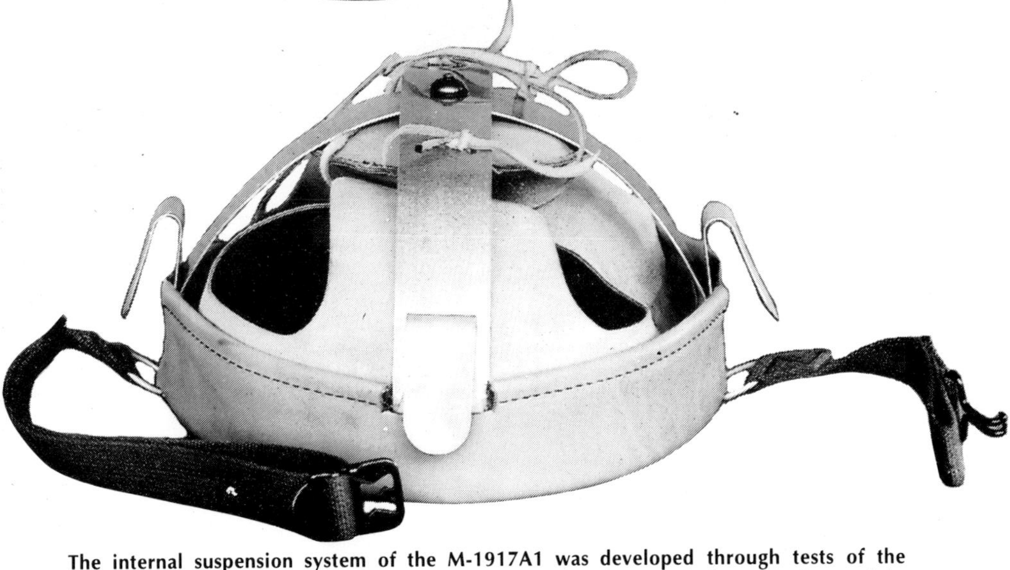

The internal suspension system of the M-1917A1 was developed through tests of the Experimental Model 5A in the 1920s and early 1930s. This photograph shows the frame of the suspension before and after being covered in its leather padding. (Photo Courtesy of Rock Island Arsenal Museum)

A close-up of one of the ends of the suspension system. These bends in the bands acted like springs and provided tension. Notice the chinstrap is attached to a chinstrap loop, secured to the liner system.

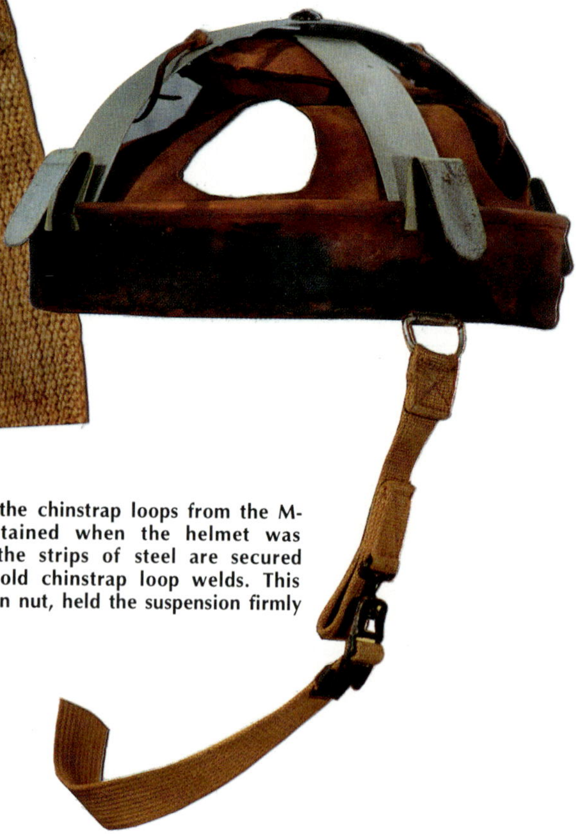

Many collectors wonder why the chinstrap loops from the M-1917 helmet shell were retained when the helmet was converted. The reason is the strips of steel are secured underneath the edge of the old chinstrap loop welds. This tension, coupled with the crown nut, held the suspension firmly inside the helmet shell.

The correct method of positioning the chinstrap on the M-1917A1. The chinstrap is snaked through the old chinstrap loop retained from the M-1917 shell.

Collectors are often at a loss to explain why the old M-1917 chinstrap bails were retained when the helmet liner was altered. The answer: The tension springs on the sides of the helmet liner system were anchored beneath the edge of the old bail which extended into the bowl of the helmet.

Inside the helmet's crown, a round four-inch diameter leather pad provided additional comfort. The pad was held in by a pair of leather laces tied to holes punched in the steel band which ran from side to side.

The steel headband which ran the circumference of the interior of the helmet was covered with a leather sweatband. This sweatband had four leather leaves which extended into the center of the helmet shell. A small steel grommet was placed at the end of each leaf. At the point where the four leaves met in the center, a leather thong tied them firmly together.

These leather leaves supported the helmet on the head and were a marked improvement over the old M-1917 liner.

Another improvement to the M-1917A1 was its chinstrap. If the chinstrap broke or wore out on a M-1917, the helmet was basically out of service, since the leather chinstrap was the liner system's main support.

On the M-1917A1, the chinstrap came in two parts and was independent of the suspension. The chinstraps were made of a thick, durable, yellowish-brown, canvas material looped over and sewn, one on each side, onto the chinstrap loops built into the steel band suspension.[2] The longer portion of the chinstrap, which held the buckle, was on the right, and the short side, with the hook was on the left. When installed, the new chinstraps were weaved through the old M-1917 chinstrap loops.

Once the new liner system and chinstrap were installed, the helmet conversion was considered complete.

---

[2] *Why the chinstraps were sewn to the chinstrap loops, rather than using an alternate method, is unknown.*

The M-1917A1 was the first American helmet to officially use the webbed chinstrap with its "hook and arrow" buckle system. More durable and practical than leather, this webbed style of chinstrap would remain in service on U.S. helmets into the 1980's. The chinstrap was threaded through the chinstrap loops and *bar tacked* (sewn) in place.

It would, however, appear that in many cases, the helmets were also repainted during this conversion process. The vast majority of M-1917A1 helmets encountered in museums or in the hands of collectors are finished in a pale green color and have a smooth exterior texture.

The original M-1917 helmet was painted olive-drab and subsequently coated with a fine dusting of sawdust.[3] Since M-1917A1 helmets were not specifically produced until 1940, it seems logical to assume that some refinishing of helmets may have also taken place when the upgrade to the M-1917A1 took place.

This photograph obtained from the Rock Island Arsenal Museum shows M-1917A1 steel shells in three different finishes or "coatings" as the caption on the photo stated. Taken in July 1935, it can be assumed the Army may have been testing various "coatings" or textures for the helmet. It is obvious the dark sawdust-coated helmet fared better than the competition. (Courtesy of the Rock Island Arsenal Museum)

[3] *See Chapter One*

Although this helmet is similar in appearance to the dark sawdust-coated helmet at lower left, it is coated in sand.

As the 1940's approached and events in Europe begin to affect military readiness in America, the Ordnance Department of the U.S. Army started to mobilize for possible involvement.

The only manufacturer of the M-1917A1 helmet was the McCord Radiator and Manufacturing Corporation in Detroit. McCord bid on, and was

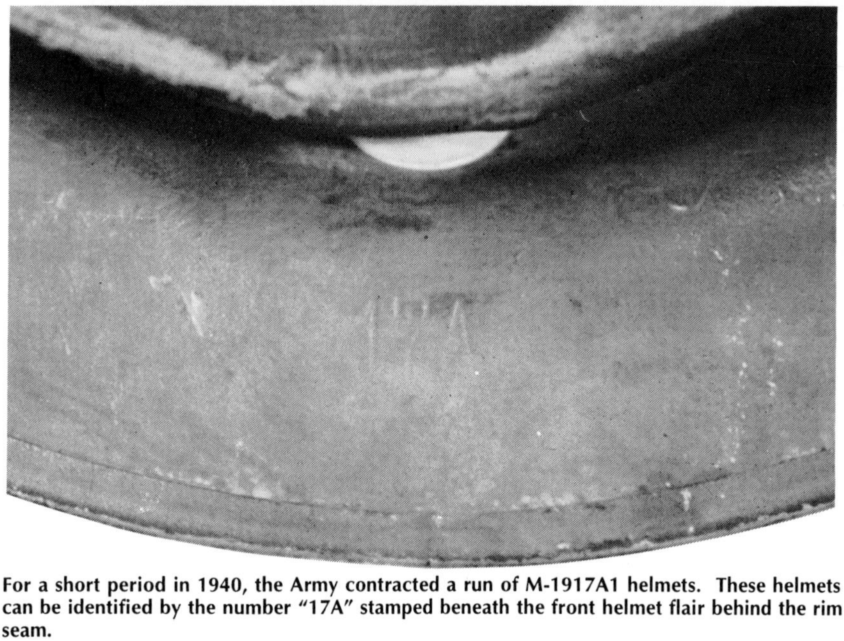

For a short period in 1940, the Army contracted a run of M-1917A1 helmets. These helmets can be identified by the number "17A" stamped beneath the front helmet flair behind the rim seam.

issued, a contract from the Rock Island Arsenal for nearly two million M-1917A1 steel helmets.

McCord, while very experienced in the manufacture of steel components, had never made helmets and consequently had to learn quickly. The company was ultimately successful and completed more than 900,000 M-1917A1 helmets.

The final chapter in the M-1917A1 story came in 1941. The original contract issued from the Rock Island Arsenal was terminated in favor of the manufacture of the next generation of U.S. combat helmets.

The M-1917A1, a throwback to World War One, would, however, be worn by every branch of service as America's primary combat helmet in the early stages of World War II.

The soldier above wears a M-1917A1 helmet while communicating with the newly-developed "walkie-talkie" radio. (Courtesy of the George A. Petersen collection)

This M-1917A1 has been painted blue for use with U.S. Naval operations. All M-1917A1 helmets were painted olive drab. Helmets painted like this example were repainted once they arrived at the ship or naval installation. (Courtesy of the David B. Powers collection)

These two Marines are believed to be members of a Marine Corps band in China in the 1930s. They are outfitted with M-1917 style helmets refinished in high gloss paint. Attaching the USMC Eagle, Globe and Anchor (EGA) insignia to the helmet was not a standard practice in the Marine Corps. While Hollywood often shows the "EGA" attached to the M-1917 style of helmet, normally only special units such as bands and honor guards, and some senior officers, attached the insignia to the helmet. While the soldiers in the photos wear what appear to be M-1917 helmets, it seems to be more common to see the EGA attached to the M-1917A1 helmet.

A M-1917A1 helmet bearing the 1937 example of the USMC Eagle, Globe, and Anchor (EGA). Helmets in various types of finish had the EGA attached.

This is a most unusual M-1917 style helmet cover consisting of a two-piece white cotton material formed into a cover with a drawstring to hold it in place. A USMC EGA printed on a separate piece was sewn onto the cover. (Courtesy of the Tim Sloat Collection)

Below: Found on an M-1917A1 helmet, this cover is made of tan cloth with foliage loops sewn in place (Courtesy of the Keith R. Jamieson, MD, Collection)

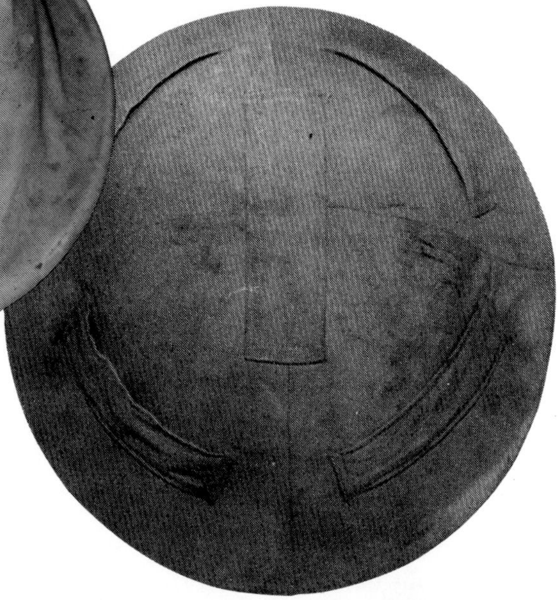

An interesting pattern camouflage cover being worn on an M-1917A1 helmet. During the late 1930's and into WWII the Army experimented with a variety of camouflage patterns and designs. Ultimately, the Army did not adopt a camouflage cover for its helmets during World War II.

CHAPTER FOUR
THE M1 HELMET OF WWII

With the 1930's about to end, America once again found itself on the sidelines watching events in Europe unfold. As the *Blitzkrieg* of Hitler's Third Reich pulverized Europe and threatened Britain, America began to prepare for war.

This war would result in the introduction of the most famous of all American combat helmets, the M1.

Since World War I, efforts at developing a distinctive American helmet had been attempted on several occasions, yet no single example evaluated had been accepted for adoption. The M-1917 helmet and its upgraded successor, the M-1917A1, were widely criticized as lacking in protection and were generally uncomfortable.

One idea put forward in the early 1930's was a two-piece combat helmet. Essentially the configuration would be a suspension system constructed of a lightweight material which would fit into a protective steel shell. The Army did conduct some initial research into the viability of this concept but determined that the helmets would be too complex to manufacture.

In August 1940, when Assistant Secretary of War, Robert P. Patterson, was requested to sign an order for additional M-1917A1 helmets...he refused! Instead, he directed that a new, improved helmet be adopted. The Ordnance Department procrastinated, and eventually ordered more of the old model helmet.

In November of that year, the Chief of Infantry, Major General George A. Lynch, also ordered that a new helmet should be designed. The assignment went to Major Harold G. Sydenham in the Infantry Board's Test Section (formerly the Department of Experiments), at Fort Benning, Georgia. Sydenham was already known for his ideas about a new helmet design, which incorporated the two-unit, shell-and-liner innovation.

After research and review of past attempts at helmet development, the board identified the elements it felt important for a new helmet. These elements addressed the following items:

protective ability
weight
comfort
balance and stability

Sydenham convinced the Infantry Board to revive the early 1930's idea of a two-piece combat helmet. These elements were compiled into a suggested specification which was submitted to the U.S. Army Ordnance Department for action.

The Ordnance Department immediately contacted The Metropolitan Museum of Art in New York. Since the Museum's Arms and Armor Department had been instrumental in designing experimental helmets during the First World War, it was hoped they would again provide expertise.

The earlier evaluation performed by the Infantry Board determined the M-1917 helmet provided superior protection for the top of the head. Yet it provided insufficient protection for the side and rear.

Another area of prime concern was to gauge the opinion of the soldiers who would be wearing the helmet. The Infantry Board wanted to leave nothing to chance and solicited input from regular GIs on everything from the helmet's shape to how comfortable it was. Experience gained during the development of helmets during World War I showed that the best helmet design in the world would not protect a soldier if he didn't wear it. As strange as it may seem, the Liberty Bell style helmet developed during World War I, while far superior in protective capabilities to the M-1917, was overwhelmingly rejected by soldiers who refused to wear it. The reason for the rejection was simple, they disliked its strange cone-shaped appearance.

Those considerations, coupled with the specifications desired by the Infantry Board, became the basis of the new helmet. They determined that combining the attributes of the M-1917's crown with the side and rear protective qualities of the experimental Model 5A could be the answer. An M-1917 helmet was altered by shearing the flairs off and then welding on steel sides and a neck skirt similar to what a bowl looked like. This operation is said to have taken place at a blacksmith or metal shop on Fort Benning under the direction of Major Sydenham. This prototype mixed the positive aspects of both the M-1917 and the Model 5A. The result was tested and eventually designated the TS-1.[1]

The TS-1 showed promise right from its inception and the Army decided to have several manufactured for extended field and ballistic testing. Another prototype, designated the TS-2, was made by armorer Leonard Heinrich at the Metropolitan Museum of Art in New York. The TS-2 was die-cast with a bolt in the crown to hold the still-to-be-developed liner in place. When it was discovered that the liners did not need to be permanently attached to the steel shell, the bolt was eliminated.

For the manufacture of the TS-3's steel shell, the Army contacted McCord Radiator and Manufacturing Corporation. McCord was the only producer of combat helmets and was currently under contract to manufacture approximately 2,000,000 M-1917A1 helmets.

Working in McCord's favor was the decision by the Army to use Hadfield Manganese as the steel for the helmet shell. This tough, hard metal

[1] *The "TS" prefix stood for Test Section and indicated the helmet was experimental.*

The original TS-1 helmet design was created when an M-1917 helmet shell was altered. Part of the flair was removed and additional steel covering the sides and rear was welded on. The helmet was then pounded into shape. The helmet in this photo is believed to be an original TS-1. Notice the weld line on the side of the helmet. (Courtesy of the David B. Powers Collection)

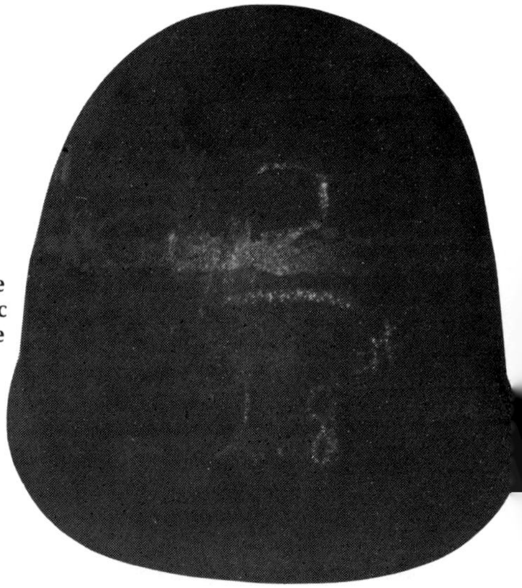

The rear view of the same helmet. Again, notice the weld line and how it conforms to the basic shape of the crown of an M-1917 helmet. The TS-1 design was an obvious improvement in protection. (Courtesy of the David B. Powers Collection)

was well suited for armor. McCord was experienced in using Hadfield Manganese because the M-1917A1 helmet was also constructed of it.

McCord's major concern was whether the seven inch deep bowl design of the TS-3 could be formed in a single drawing operation. McCord believed the operation could be done, but the steel had to be prepared very carefully prior to being pressed.

McCord's die engineers went to work designing the dies to press the shape of the TS-3.

At the same time, metallurgists determined the steel had to be rolled into very precise, consistent sheets if it was to be formed into the shape of the helmet without heating the steel during the operation.

Initially the failure rate of the steel shell was as high as fifty percent but as improvements were made, the rate fell to below 10 percent.

After a number of the shells were produced, ballistic tests showed the helmet exceeded the protective qualities required by the helmet specifications. Like the M-1917, the TS-3 was tested by firing a .45 caliber bullet at the helmet shell. While the M-1917 was required to withstand the impact of a

.45 bullet at 600 feet per second, the TS-3 had to perform against a bullet traveling 800 feet per second.

It did so quite successfully.

The new helmet was approved on 9 June 1941 and was officially christened the M1. Sydenham was half-way toward completion of America's new combat helmet!

This soldier wears an early example of the refined TS-3. Notice the round rivet securing the chinstrap to the steel helmet shell. The liner used in this helmet shell appears to be a Hood Rubber liner. (US Army Photo)

Developing the M1 Liner

With success apparently met in the area of the steel helmet shell, Sydenham's focus changed to acquiring an appropriate liner to complete the helmet.

The key element of the liner system was the development of both a liner shell as well as a suspension system to install in the shell. To further complicate matters the Army wanted both the liners and helmet shells to be independent, yet interchangeable items. In other words, any liner should snugly fit any steel shell.

Both suspension and the liner owe much to a design provided by John T. Riddell. In 1941 Riddell owned the Riddell Company, a firm which manufactured football helmets (and still does!).

John Riddell, whose company made protective helmets in the 1930's played an important role in the development of the M1 helmet liner. General George S. Patton, inspired by helmets worn by the Green Bay Packers, contacted Riddell and asked him to provide some examples of the football helmet for evaluation as a possible tanker's helmet. While at Fort Benning, Riddell apparently struck up a conversation with Major Harold Sydenham, who incidentally was working to develop a new infantry helmet. (Provided courtesy of Robert Southee)

At the request of General George S. Patton, Riddell contacted the U.S. Army about equipping tanker troops with football-style helmets. The officer he spoke with was Colonel Sydenham.

Their conversation moved from tanker's football helmets to the feasibility of adapting a football-helmet-type suspension into a steel combat helmet shell.

History indicates Sydenham and his wife, Zelda, actually formed an example of the first TS-1 helmet liner out of a plastic compound called Vinylite in their kitchen at Fort Benning. It should be noted that John T. Riddell's suspension system was not installed in the original liner, nor did he participate in creating the original prototype. But Riddell's son, Jack, later worked with Sydenham and the Test Section to incorporate a Riddell suspension system into the design.

Very quickly after that meeting, at Sydenham's urging, the Army and Riddell signed a licensing agreement. This agreement allowed the Army to use a suspension system, similar to that used in Riddell's football helmets, in the new helmet liner.

After the crude TS-1 liner was created by the Infantry Board's Test Section at Fort Benning, an interim prototype was made for die-casting by armorer, Leonard Heinrich, at the Metropolitan Museum of Art in New York, resulting in the TS-2. Because Sydenham believed it would be necessary to attach the liner to the steel shell, the TS-2 had a bolt on the top.

Initially attempts were made to construct the liner using injection molded plastics, but the resulting product was much heavier than the Army could allow. Thus the search was on for an alternative to molded injected plastics.

Back in Detroit, McCord had been advised that the contract for all remaining M-1917A1 helmets would soon be terminated. With the dies for the M1 steel helmet shell almost perfected, the need for liners became urgent.

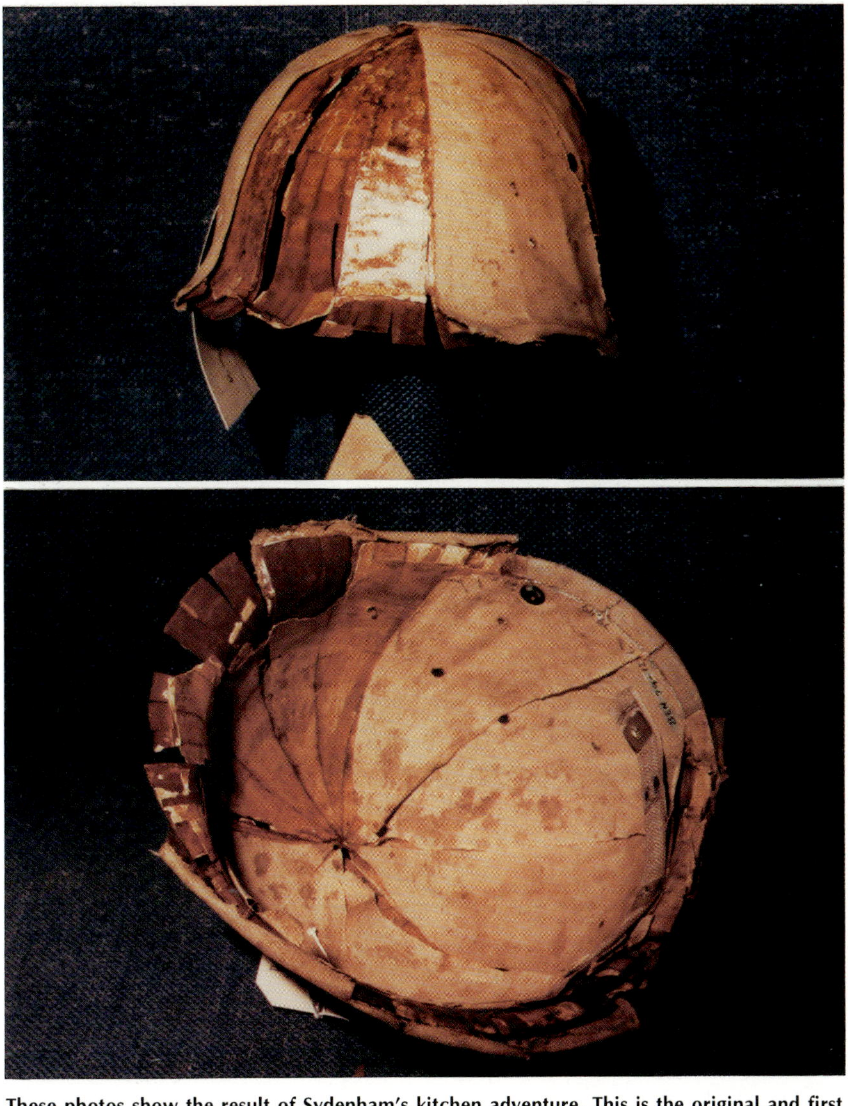

These photos show the result of Sydenham's kitchen adventure. This is the original and first liner for what would soon be the M1 helmet. Obviously the liner is showing its age. The number at the inside bottom appears to be a date of 6 December, 1979. This is probably the date the liner was cataloged into the museum at Fort Benning. (Photo courtesy of Robert Southee)

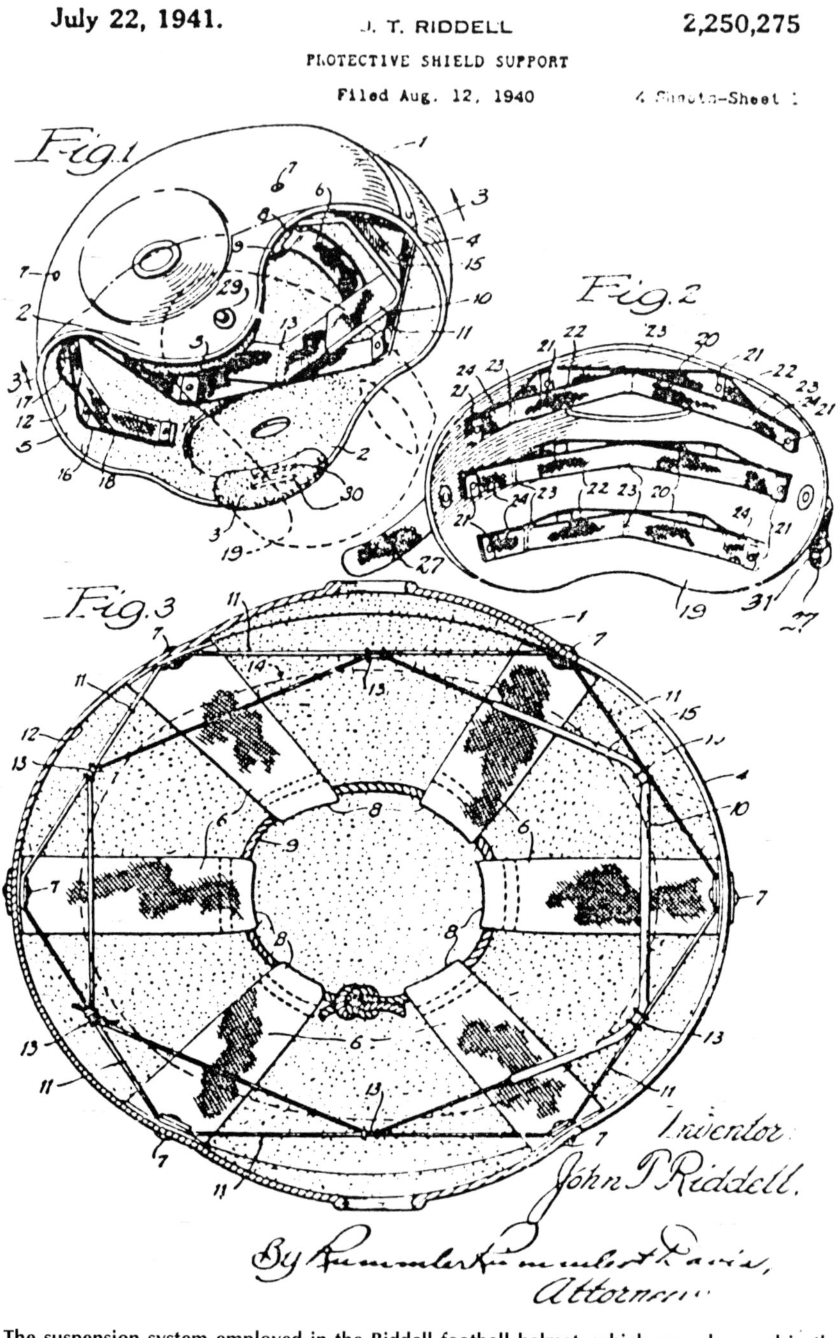

The suspension system employed in the Riddell football helmet, which was also used in the M1 liner.

McCord had been advised that the contract for helmet production would be awarded to them, yet they were unable to proceed in production without suitable liners.

Boldly taking the initiative, McCord contacted Hawley Products Company in St. Charles, Illinois. Hawley was, at the time, the manufacturer of the Army's pressed paper, tropical helmet.

McCord, without the consent of the Army, proposed to Hawley, that if they were able to construct some samples of a liner which could fit into the steel shell -- plus develop the dies to make that shell -- Hawley would be in a very good position to sub-contract the product from McCord.

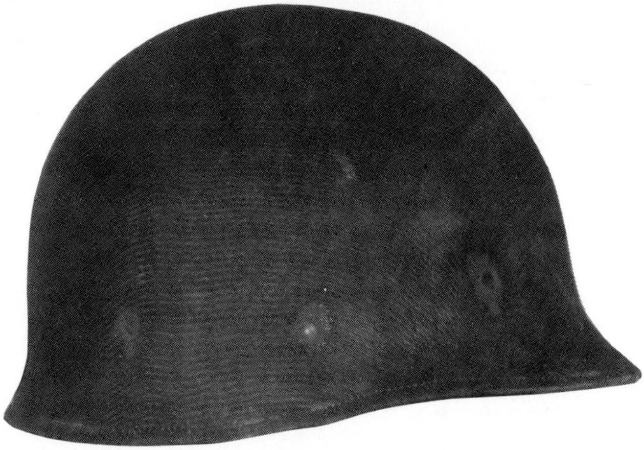

As the M1 helmet project was nearing contract stage McCord Radiator asked the Hawley Products Company to develop a liner. This photo is believed to be of an early Hawley prototype for a liner. The twill covering material and thick body, and chinstrap are very similar to production Hawley liners. (Courtesy of the David B. Powers Collection)

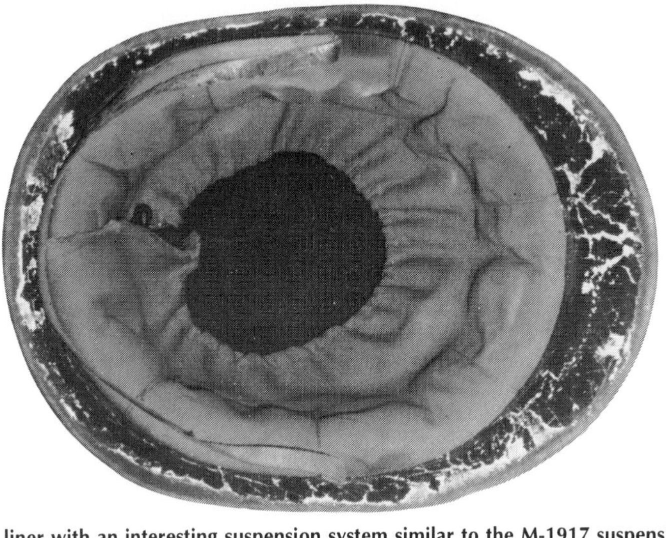

The same liner with an interesting suspension system similar to the M-1917 suspension. Like the M-1917, the crown of the liner is padded with a gray felt-like material. The suspension is a khaki twill and incorporates a lace to adjust the suspension. (Courtesy of the David B. Powers Collection)

After perfecting the product, Hawley products provided this sample helmet liner to Major Sydenham to present to the Infantry Board for consideration. (Note the initial I.B. for Infantry Board). This helmet was worn by Sydenham after being promoted to colonel. (Photo courtesy of Robert Southee)

Hawley decided to accept McCord's offer and very quickly designed and manufactured a shell which would accommodate the Riddell style suspension and fit snugly into the steel shell of the M1.

This is all the more remarkable considering no specifications, or even an official Army description, for the liner existed!

Nevertheless, with samples in hand, the McCord Company presented 100 examples of Hawley's work to Sydenham to present to the Army for testing. It was soon discovered that these liners stayed firmly in place without a bolt, and it was eliminated in the TS-3 and subsequently the M1. Suitably impressed, the Army notified both McCord and Hawley that the liner was satisfactory and asked that the company prepare for orders. Sydenham's quest to develop a distinctive American helmet was complete.

Feb. 29, 1944. H. G. SYDENHAM Des. 137,390

HELMET

Filed April 7, 1943

Fig. 1

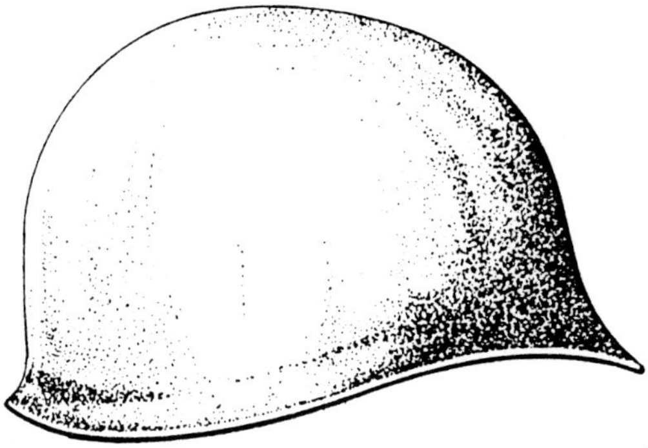

Fig. 2

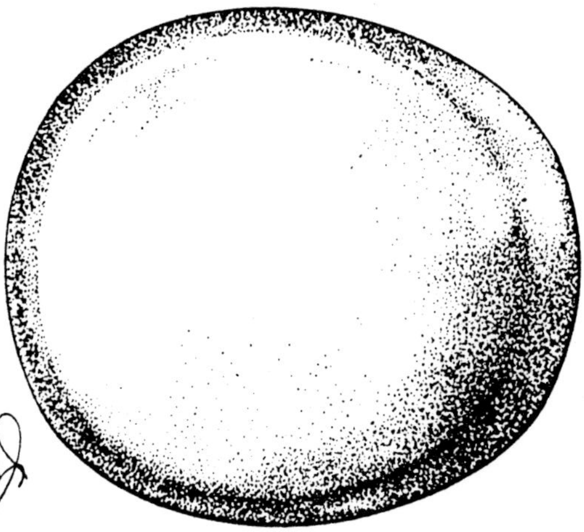

INVENTOR.

H. G. Sydenham

BY Robt. Williams Jr.

ATTORNEY

UNITED STATES PATENT OFFICE

137,390

DESIGN FOR A HELMET

Harold G. Sydenham, United States Army, Fort Benning, Ga.

Application April 7, 1943, Serial No. 109,918

Term of patent 14 years

To all whom it may concern:

Be it known that I, H. G. Sydenham, United States Army, a citizen of the United States, residing at Fort Benning, Georgia, have invented a new, original, and ornamental Design for a Helmet, of which the following is a specification, reference being had to the accompanying drawing, forming part thereof.

Fig. 1 is a side elevational view, and

Fig. 2 is a top plan view thereof.

I claim:

The ornamental design for a helmet, as shown.

HAROLD G. SYDENHAM.

The original drawing and patent submitted by Harold G. Sydenham, the designer of the M1 helmet. It is interesting to note that the actual submission date was nearly two years after the M1 helmet was adopted. (Provided courtesy of Robert Southee)

Forming the M1 Steel Shell

Back at McCord in Detroit, the M1 production line had been placed into action and helmets were soon flowing. Before the end of the war the complete manufacturing process from pressing to packaging a helmet shell for shipping, consisted of 27 separate steps taking 22 minutes.

The process for construction of the M1 helmet was very similar to that of the M-1917.

The steel was carried to Detroit on flatbed railroad cars. Upon arrival at the factory, the steel was identified and marked.[2] As the steel entered the McCord factory it was sent directly to presses which formed the helmet shells. The press stamped out two helmet shells at a time.

After that pressing operation, the helmets were run through a second press. This smaller unit had the task of cleaning up and trimming off the sharp and uneven edges that resulting from the main press.

After trimming, the front visor section of the M1 helmet was formed. Rather than being stamped into shell, the front of the helmet was actually bent into shape.

The next step was applying the rim edging. The helmet rim was positioned around the perimeter of the helmet then spot-welded into place.[3] Shortly after the rim was positioned, the chinstrap loops were attached. The early versions of the M1 loops were simply loops of steel welded onto the inside of the helmet shell.[4] These welded-on loops were replaced by hinge style, or swivel loops, in late 1943.

It was discovered that the paint did not adhere to the stainless steel used on the helmet rim. Stainless steel was used for the rim due to its resistance to rust and non-magnetic qualities. Unfortunately, when the paint chipped off the rim, the stainless steel shined brightly. This resulted in the rim material being changed to steel made from Hadfield Manganese in 1944.[5]

After the rim and the loops were attached, the helmet was rinsed in a phosphate bath to clean and score the surface to help the paint adhere to it.

As with the M-1917, the painting process was completely automated, with the helmets being carried to the painting area on a conveyor. The M1 shells were first sprayed in a dark olive-drab. The paint was mixed with cork[6] to provide texture and to break up glare. It also helped to deaden the sound of metal objects striking the shell.

After painting, heavy cotton chinstraps were sewn onto each of the helmet loops. The chinstrap on the left side of the helmet was a short, non-adjustable strap with a wire hook. The strap on the right side was much longer

---

2 As with the M1917 during World War I, this mark, a set of numbers, signified the heat of the steel to track the steel through the manufacturing process.

3 There is considerable controversy among M1 helmet collectors about the position of the seam where the helmet rim joins. Although unable to locate any data from the period, I am convinced that the seam migrated from the front of the helmet to the rear in 1944 when the helmet rim material was changed.

4 These steel loops, often called "fixed loops" were very fragile and breakage was common if the helmet was dropped on a hard surface.

5 There was a concern that the stainless steel could effect compass accuracy.

6 All M1 helmets produced in WWII were textured with cork. Sand or sawdust was not used.

and could be adjusted. The wearer positioned this strap underneath the chin and buckled the strap to the short end. Both the brass buckle and hook were coated with a chemical oxidation prohibitive agent which turned them black.

An early example of the M1 helmet mated with a Hawley fiber liner. Notice the shine to the rim seam. Early production M1 helmets were constructed with stainless steel rims. The helmets markings indicate the owner was assigned to the Army Exchange Service.

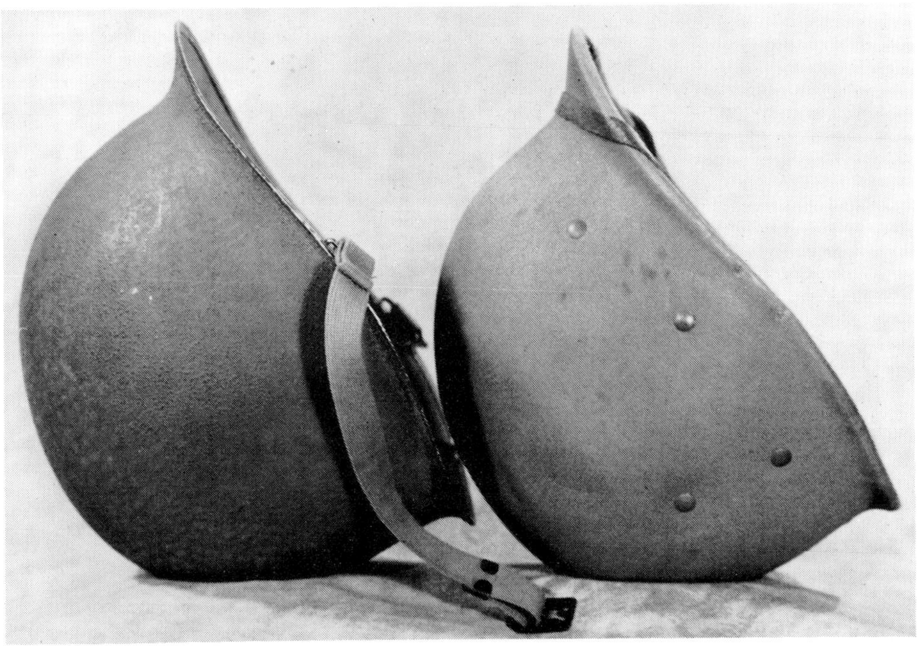

The steel shell (left) and the thick Hawley fiber liner. This two part helmet system was unique among military helmet designs during the Second World War.

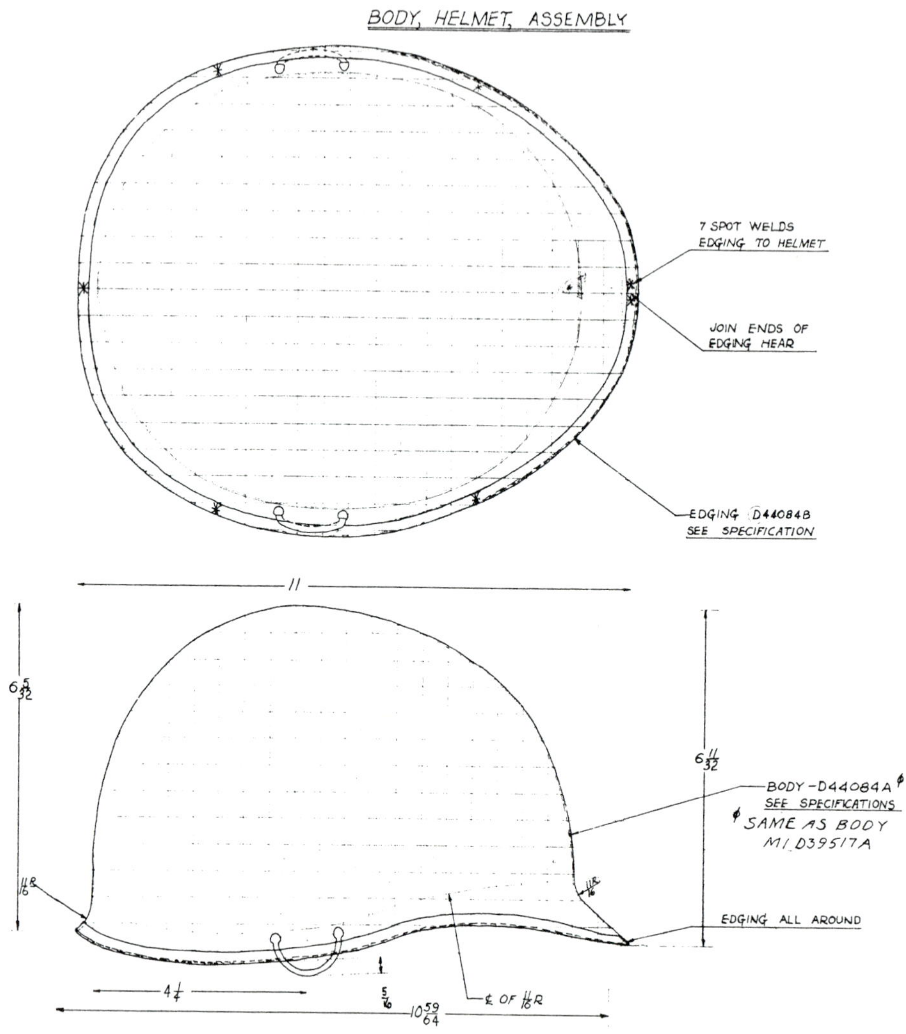

The texture of the M1 helmet shell was created by mixing cork with the paint. On early production M1s, the rim was joined at the front. While the actual date is unknown, it is probable that the seam "migrated" to the rear of the helmet in late 1944. The leather chinstrap is typically secured above the helmet front flair.

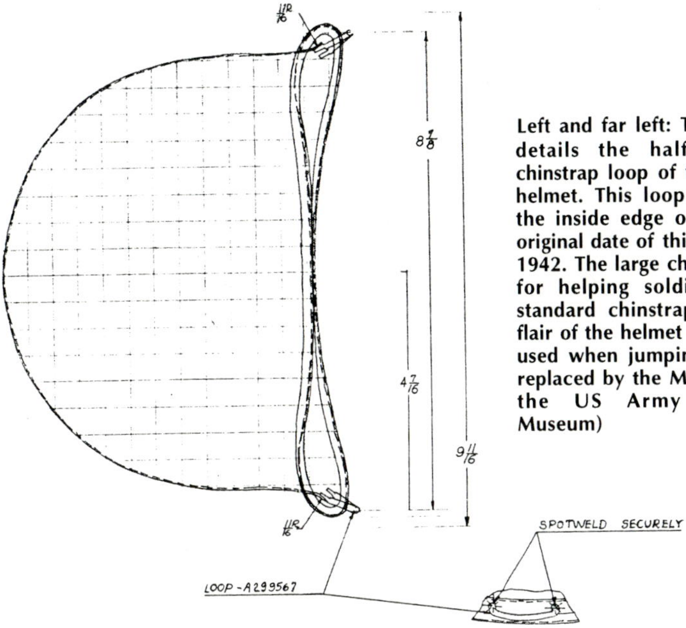

Left and far left: This specification details the half-moon shaped chinstrap loop of the M2 airborne helmet. This loop was welded to the inside edge of the shell. The original date of this specification is 1942. The large chinstrap loop was for helping soldiers secure the standard chinstrap over the rear flair of the helmet since it was not used when jumping. The M2 was replaced by the M1C. (Courtesy of the US Army Quartermaster Museum)

The welded chinstrap loop of the M2 steel helmet designed specifically for airborne forces. This larger loop was designed to aid the paratrooper in securing the steel shell's chinstrap over the back flair of the helmet when jumping. During jumps, paratroopers relied on a chinstrap system contained within the airborne helmet liner. (Courtesy of George A. Petersen)

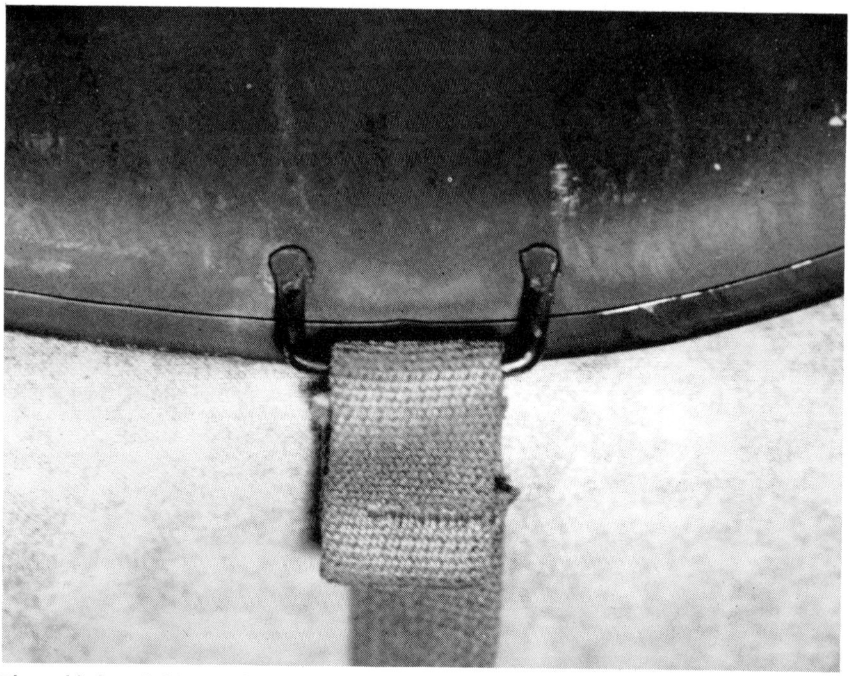

The welded steel chinstrap loop used on early models of the M1 helmet. The chinstrap was threaded through and sewn on. These loops were prone to breaking if the helmet was dropped. Made from 1941 to 1943. (Author's Collection)

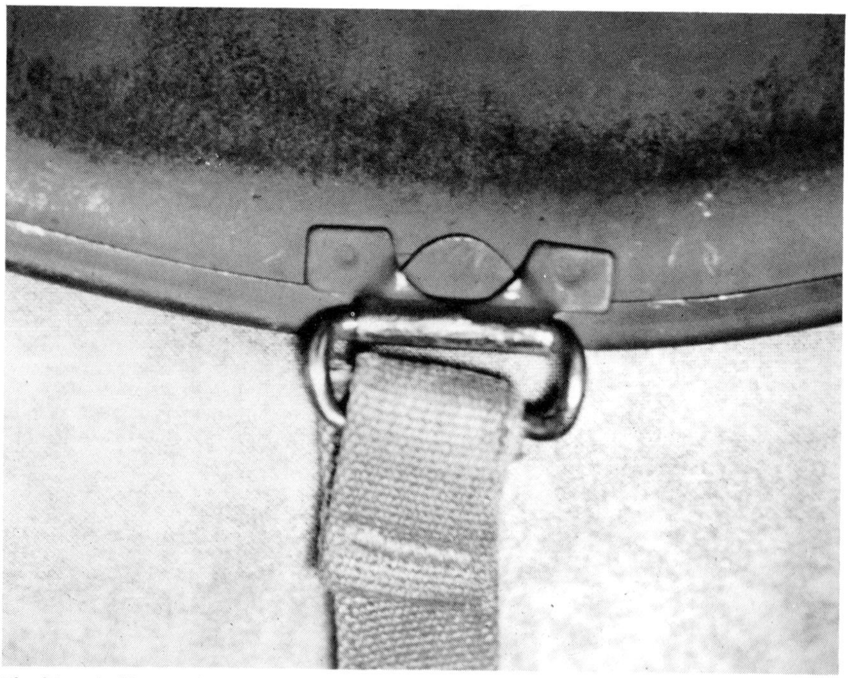

The hinged chinstrap loop replaced the welded loop in approximately October, 1943. Like the welded loop, the chinstrap was threaded through and *bar tacked* in place. This style of loop became standard for all M1 helmets to follow.

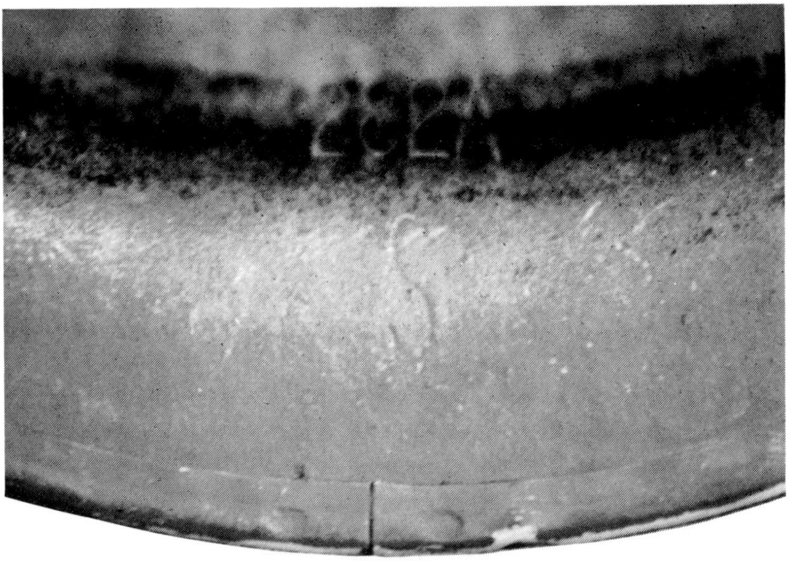

Usually visible beneath the front helmet flair are the "heat of the steel" numbers. These were used to track helmets in the event a run of steel was found to be defective. In the case of defects, helmets made from that lot could be identified and removed from the production line. The "S" marking indicates this helmet is one of two million made by Schlueter Manufacturing of St. Louis, Missouri.

Hawley's Fiber Liner

After working closely with McCord to develop the M1 liner, Hawley Products Company was given the green light to proceed with production by the Chicago Quartermaster Depot, now charged with procuring helmet liners.

Briefly, the specifications required a two-part fiber material be formed into the shape of the liner. The material would then be impregnated with varnish, to make it waterproof. After the shell was formed, an OD twill material was glued over it.

Into the interior of the helmet liner went the framework for the Army version of Riddell's suspension.

The crown of the suspension frame was initially constructed of silver colored rayon webbing and was attached to the helmet shell using washers riveted at six locations around the perimeter of the helmet liner shell. At the rear of the liner shell was a second part of the suspension which supported the liner on the nape of the neck.

Another item attached to the helmet liner was an adjustable leather chinstrap. This brown leather strap was riveted, one end on each side, to the inside of the liner. These internal attachments accounted for the 11 rivet heads visible from the outside of the liner shell.

When these liners were issued to soldiers, the adjustment bands to complete the helmet liner's suspensions were also issued. The Riddell suspension consisted of a snap-in sweatband and snap-in neck adjustment. The sweat-

The Hawley fiber liner was the first production liner. It was very light and fit snugly into the helmet shell. Hawley liners are easy to distinguish due to the thickness of the liner rim. Fragile and prone to water damage, they are becoming difficult to find.

Hawley liners are very distinctive from later plastic liners due to the construction. Hawley liners were quite fragile and susceptible to damage from moisture. After 4,000,000 liners were made the Army declared them unsuitable for service. Consequently, no more were produced. Notice how the headband is attached to the suspension with snaps. Also, only the front half of the headband is covered in leather.

The fray in the fabric of this Hawley liner is a testament to how fragile they were. Again, note how the headband snaps in place.

Notice the leather chinstrap connector. Early Hawley liners had riveted chinstrap connectors. If these broke or were damaged they could not be easily replaced or repaired.

band was made in 13 sizes and the neck adjustment in three. Like the suspension, these items were also constructed of a silver rayon webbing.

During the course of production the Hawley liner underwent several changes. First, the silver rayon webbing gave way to a standard khaki color. Secondly, the multi-sized, snap-in sweatband was eliminated in favor of an adjustable clip-in style. Finally, riveting the leather chinstrap directly to the shell was discontinued and in its place a detachable system was installed.

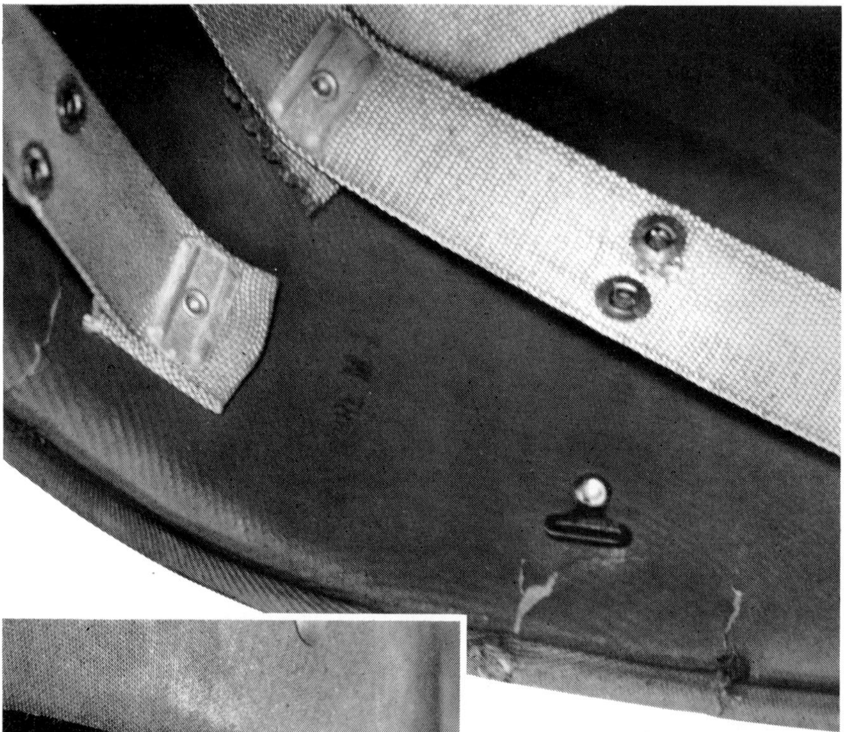

Above: "FIBRE M-1." This mark is often found stamped in ink inside a Hawley liner. Notice the snaps for the headband, the use of the rectangular washer, and the riveted chinstrap connector.

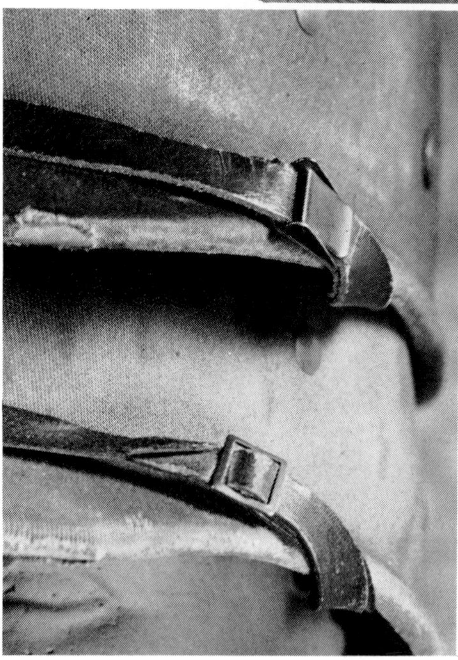

Left: A comparison between the first pattern (bottom) and second pattern chinstrap adjustment buckles on the Hawley liner. (Leffel Collection)

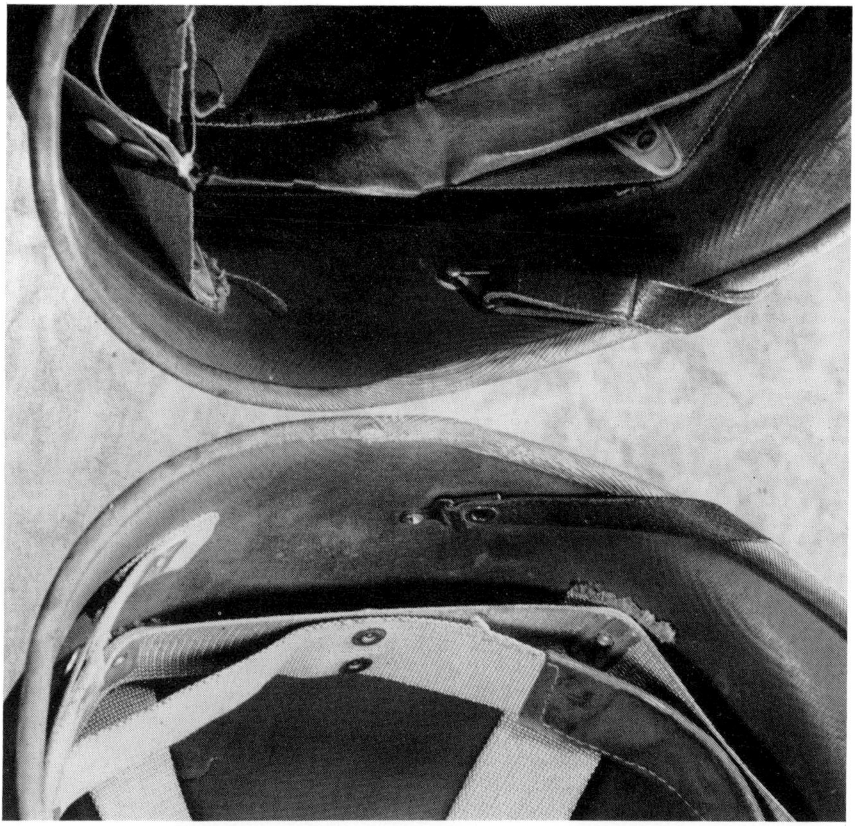

A comparison of early (left) and late-model Hawley liners. During production, upgrades were made to the suspension material. It changed from silver/gray rayon to khaki herringbone twill cotton. The chinstrap rivet also changed to a snap-on lug. The washer became "A" shaped. The most important change was eliminating the 13 individual sizes of headband in favor of an adjustable one-size-fits-all headband. (Leffel Collection)

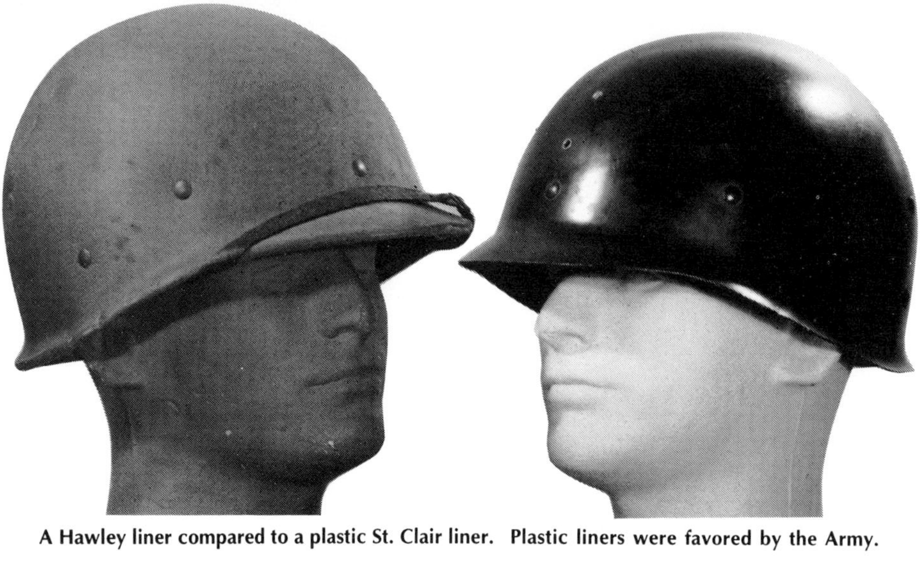

A Hawley liner compared to a plastic St. Clair liner. Plastic liners were favored by the Army.

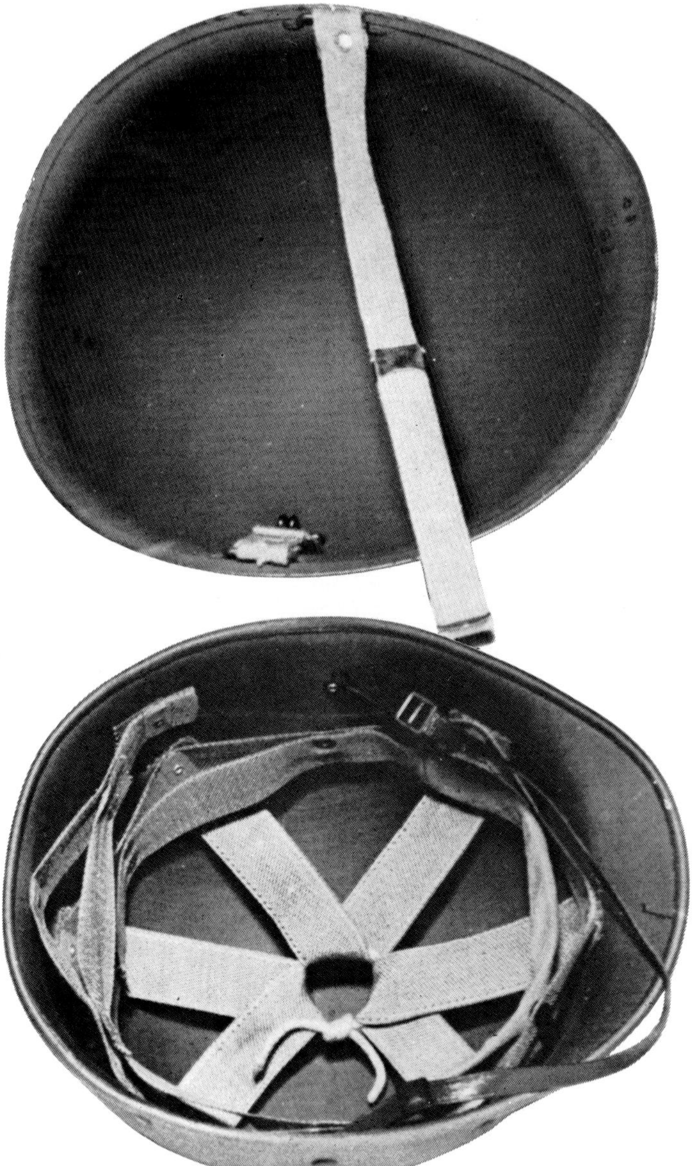

An interior look at the new TS-3 helmet designed to replace the aging M-1917A1. The chinstrap on this helmet is attached to the chinstrap loop with a rivet, instead of bar-tacking it in place. The liner is, of course, a Hawley-made fiber liner using a modified version of John Riddell's helmet suspension system.

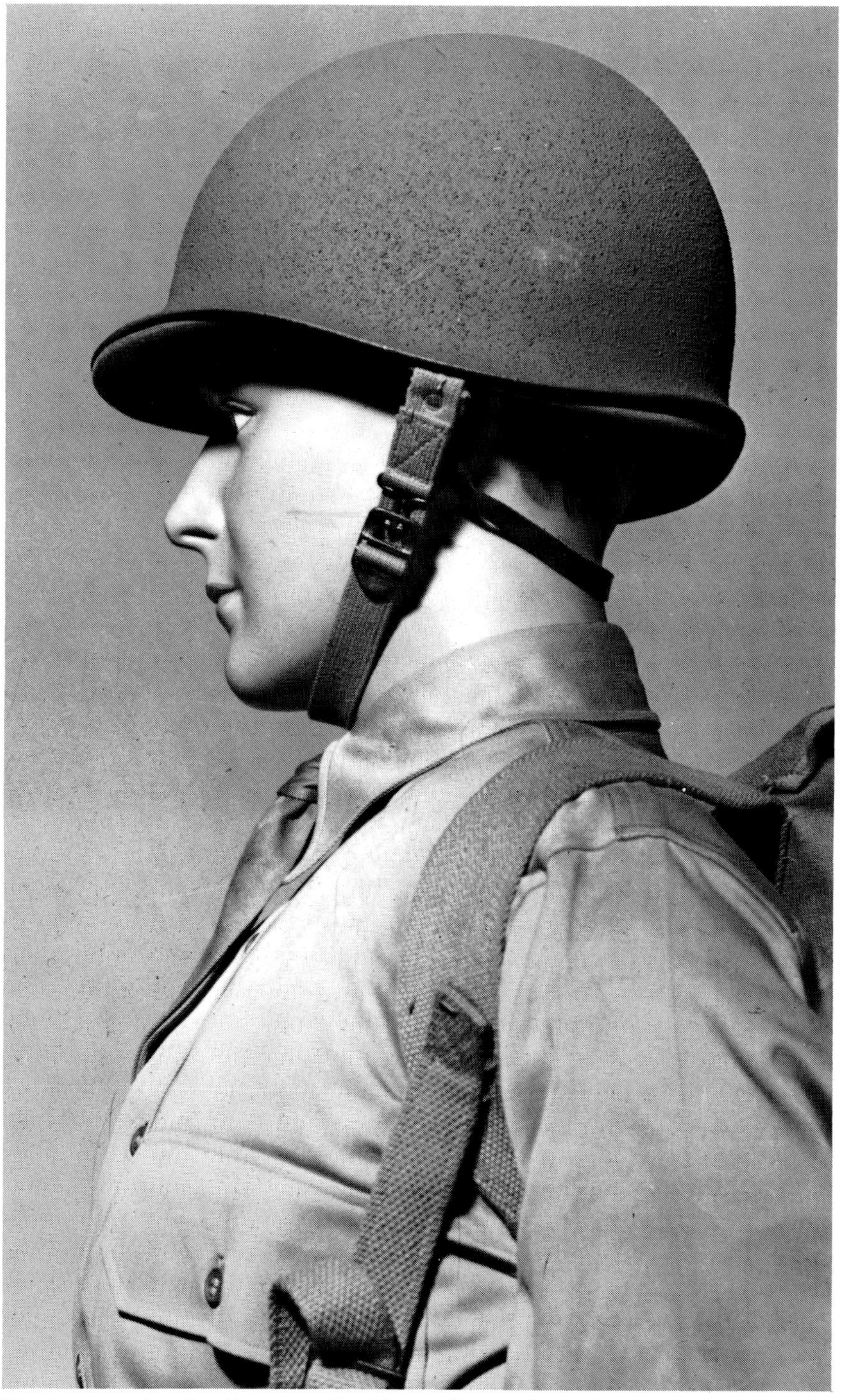

The rivets securing the chinstrap to the chinstrap loops indicate this is an experimental TS-3 helmet. Notice how the leather liner chinstrap has been positioned behind the nape of the neck rather than across the brim as normally seen. It is interesting to think this may have been the liner chinstrap's original purpose.

Another view of a TS-3. The thickness of the helmet liner visible under the steel shell indicates this may be a very early plastic liner rather than a Hawley fiber liner.

The Plastic Liner

Even as the Hawley liners were rolling off the assembly conveyors, the Army was searching for alternatives. While the style of the liner was excellent, many of the materials used to make the helmet liner were less than satisfactory. The twill covering frayed easily, the rayon suspension stretched out of shape and the helmet shell became very soggy when wet.

In hopes of being awarded an Army contract to make helmet liners, more then a half-dozen companies submitted samples for consideration. While the materials used by the companies differed greatly, one common denominator was that each company impregnated its helmet material with some form of plastic resin.

Using plastic resin in the liner was evaluated by the Quartermaster's Standardization Branch to determine which method was best. The evaluation determined that duckcloth[7] strips, soaked in a solution of resin including a small amount of phenolformaldehyde, would create a strong waterproof material. This strong plastic material could be molded into an ideal helmet liner.

[7] *A cotton or linen cloth-like canvas but finer and lighter in weight.*

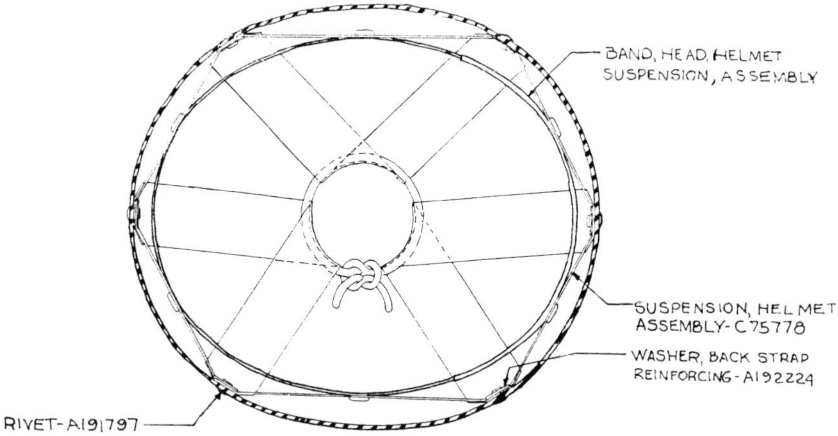

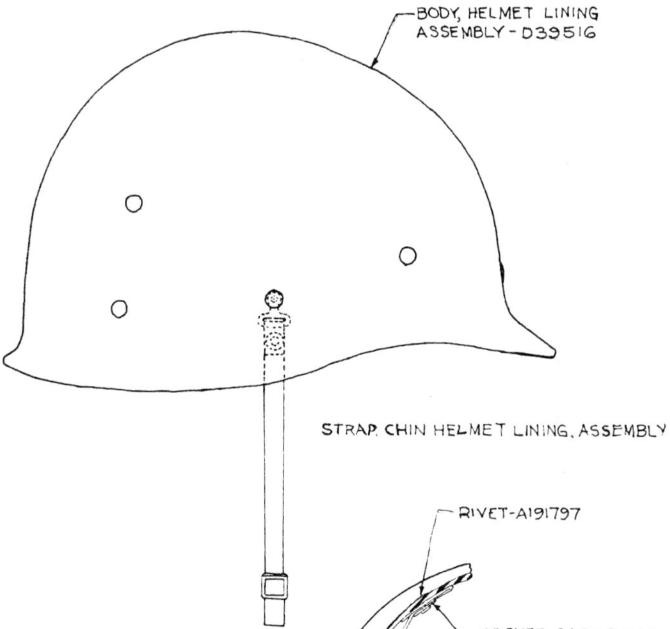

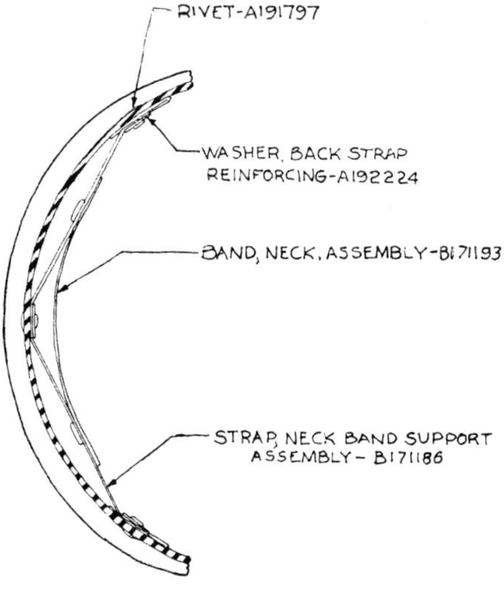

A good view of the assembly and nomenclature of the early WWII M1 liner. Notice the permanent chinstrap rivet and the early style of chinstrap adjustment buckle. This early specification details the snap-in headband and neck band. On the standard M1 liner shell there were 11 rivets (plus the insignia eyelet). WWII airborne liners have two additional rivets for the female snap fixtures, for a total of 13 rivets plus an insignia eyelet. Hawley fiber liners will not have an insignia eyelet. This diagram is dated 15 October 1941. (Courtesy of the U.S. Army Quartermaster Museum)

Six hundred initial test samples of the plastic M1 liner were made by General Motors' Inland Manufacturing Division. Broken into three lots of 200, the liners represented three different chemical formulas from three different companies.[8] Each company did, however, use a high-pressure method of construction.

The plastic liners were dispatched for testing to Aberdeen Proving Ground, where they were joined by samples provided by the St. Clair Rubber Company which used a low-pressure method to form its helmet liners.

The test results showed the liner formula designed by Westinghouse performed the best, with the Hawley liner falling behind all the plastic versions in durability and protective qualities.

Those factors caused the Army to quickly adopt the plastic liner made from duckcloth impregnated with resin as the standard for liners for the new M1 helmet.

While the Army would have been very pleased to require each manufacture to use the specific formula developed by Westinghouse, the urgency of acquiring liners prohibited that scope of standardization. Consequently, the initial specifications for the plastic M1 helmet liners were vague.

Initially, a high degree of tolerance for product variability was permitted to expedite liner manufacture.

Contracts for plastic liner manufacture were awarded to 10 companies:[9]

- Inland Manufacturing Division of General Motors Corporation for one million.
- Westinghouse Electric Company for one million.
- Mine Safety Appliance Company for 384,000.
- St. Clair Rubber Company for 450,000.
- Capac manufacturing Company for 600,000.
- Firestone Tire & Rubber Company, Inc., for 750,000.
- Hood Rubber Company, Inc., for 200,000.
- International Molded Plastics, Inc., for 160,000.
- Seaman Paper Company for 400,000.
- Hickok Manufacturing Company, Inc.[10]

The St. Clair Liner

While the majority of manufacturers contracted to make helmet liners opted for a high-pressure method of construction, the St. Clair Company took another route.

The St. Clair liner was made using a three-step process. First, six strips of standard eight-ounce duckcloth soaked in resin were criss-crossed over a liner shell form, heated, and glued together. Then the now joined strips were placed in a die and molded into the shape of the liner using steam pressure for

---

8 The companies included Inland, Mine Safety Appliance Company and Westinghouse Electric Company

9 The average cost for the plastic liners was approximately two dollars, complete with suspension

10 Hickok Manufacturing Company, Inc., asked to have its contract canceled prior to making any liners. The request was approved and Hickok never made any M1 liners.

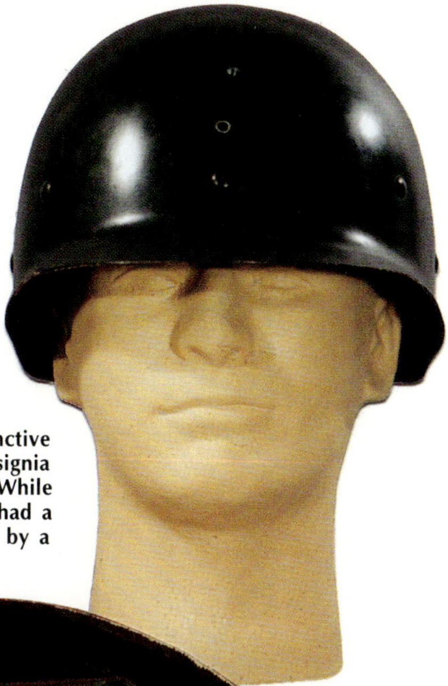

The St. Clair low-pressure plastic liner had a distinctive plastic shine. An addition to the liner was an insignia eyelet in the center of the forehead of the shell. While more durable than Hawley liners, St. Clair liners had a tendency to shatter into sharp pieces when struck by a projectile.

The interior of the St. Clair liner with its pebbly textured, dark-chocolate colored interior. Although the early riveted leather chinstrap is missing, the chinstrap rivet remains. Notice how different-colored suspension parts were intermingled. Apparently no snap-in headbands were made in khaki.

10 minutes. The final step included cleaning up and sealing the frayed edge, applying a sealer and baking on the finish.

At first glance, externally, the St. Clair low-pressure liners appeared very similar to liners made using other manufacturing methods. However, they do have two specific characteristics: Their finish appears much smoother and darker, giving it a very plastic appearance; the sides of the liner were slightly indented to accommodate the welded chinstrap loops on the M1 helmet. This feature is absent from all other M1 liners. The interior of the helmet liner is a chocolate brown and has a pebbly appearance. In the crown are the stylized letters SC applied in yellow paint.

There were several drawbacks to St. Clair's low-pressure method of construction. When the liners were being formed using the steam pressure, they were required to cure for 10 to 15 minutes. This was a substantial drag on the process, as demand for the liners was extremely high.

A more disturbing fault was a demonstrated tendency to fragment into dangerous shards of plastic during ballistics tests at Aberdeen Proving Ground.

The Hood Rubber Company Liner

Like the St. Clair Company, the Hood Rubber Company choose a method other than high-pressure to make its version of the M1 liner.

The process used by Hood Rubber is called the Ball-Winding Technique.

Considered more of a medium-pressure method, the ball-winding technique was very innovative and required 15 to 20 percent less material then other liner manufacturing processes.

Hood's method of construction was unique. Long ribbons of cotton sheeting were impregnated with resin and then cut into long strips, some two inches wide and some three-quarter-inches wide. After being cut, both sizes of strips were wound onto spools.

The interior of the Hood Rubber Company liner. These can be distinguished by the 3 1/2-inch diameter disk of duckcloth glued into the crown. Like the St. Clair liner, the Hood liner was made with a low-pressure method. (Author's Collection)

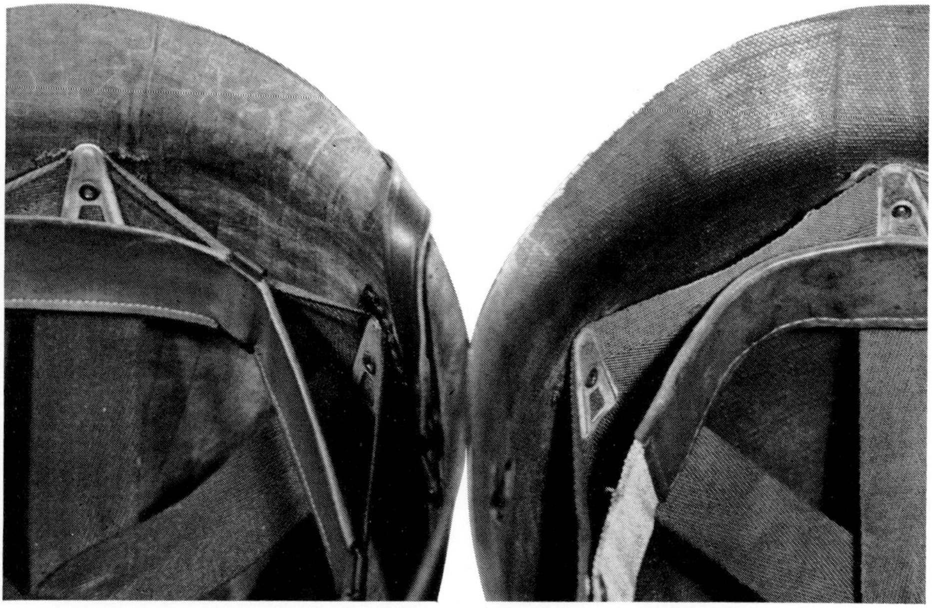

A comparison between the St. Clair liner (right) and the Hood Rubber Company liner, (left). Notice the Hood liner has the "new" clip-in headband rather than snaps.

Rather than lay out a pattern and press the impregnated cloth directly into a mold, Hood added a step and used an interesting two-piece watermelon-shaped form.

The strips of cloth from the spools were wound around the metal form. First the two-inch-wide strips and then the three-quarter-inch strips.

What resulted was a plastic-impregnated cocoon which, when cut in half, resembled two helmet liner bodies.

After being cut, the two forms had pre-molded brims glued to the shell. A three-and-one-half-inch disk of duck material soaked in resin was placed inside the crown of the liner to provide additional strength.

The liner's final shape was achieved when the forms were forced into a mold to assure the correct dimensions. Water under a pressure of 250 PSI was used during the six-minute curing operation.

After final molding, the liner was trimmed of excess material, had holes punched through it to accommodate suspension parts, and was painted in a non-reflective olive-drab.

The interior of the Hood Rubber liner is very different from the St. Clair liner and the high-pressure liners. It is a light brown and the texture gives it a very canvas-like appearance. The dozens of strips of impregnated cloth wound over the helmet form are easily seen. The three-and-one-half inch reinforcing patch in the crown of the helmet make the Hood Rubber Liner one of the easiest liners to identify. Hood Rubber Company stamped the letters HR in silver paint in the crown of the helmet -- although they appear to read 11R.

A valiant effort by the Hood Rubber Company was brought to a close by the Army after only 206,000 liners were made.

The primary reason was a small crack which appeared where the center crown patch was. While tests showed that the crack did not affect the liner's protective qualities or overall performance, it was deemed as unacceptable.

That, coupled with the Army's desire for liners made using the high pressure method, ended Hood Rubber Company's venture into the military helmet liner business.

The High Pressure Liner

Liners constructed using high-pressure methods were considered by the Army to be the best liners for the nation's combat forces.

Unlike the low-pressure method and the ball-winding technique, high-pressure liners were formed using hydraulic presses which shaped the phenolic resin-soaked duckcloth in a mold.

The Westinghouse Company manufactured its own material and resin. Other companies using the high-pressure method obtained materials from outside sources.

Step One in forming a helmet liner using the high-pressure method was to saturate the duckcloth with resin. The duckcloth was then cut into strips[11] and laid out in a pattern for pressing.

The duckcloth pattern was placed into a mold and the hydraulic press went into action, applying 150 tons of pressure for two minutes at approximately 220 degrees.

The result was a very hard, smooth surface.

An example of a high-pressure plastic liner made by the Mine Safety Appliance Company. This liner uses a mixture of old and new suspension materials. The Army allowed makers to use up existing stocks of silver/gray rayon prior to switching to khaki.

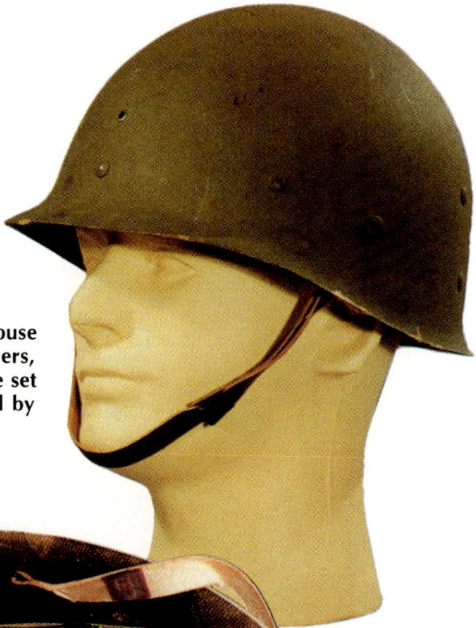

A plastic helmet liner made by the Westinghouse Corporation. Of the nine companies making liners, six used the high pressure method. Westinghouse set the standard, the formula they used was adopted by several other companies.

The preferred plastic helmet liner. Made by Westinghouse, these liners were highly desired by the Army. Shown is a very common configuration: Full khaki suspension, goldish-brown striped duckcloth interior, clip-in adjustable headband, snap-in neckband, and stud and garter secured chinstrap.

Remnants of cloth and resin were trimmed off and the liners went to the final stages to prepare them for assembly.

First, a drill press punched 11 holes in the shell to accommodate all the rivets required to support the liner's internal suspension system. One of the holes was located directly in the front of the liner approximately 2 1/2 inches

---

[11] *While there were variations between manufactures using the high-pressure method most companies cut the strips into triangular shaped wedges.*

The impregnated duck has been die-cut into sectioned discs which in this operation are being stamped to produce a hat-shape. (Photo courtesy of the U.S. Army Quartermaster Museum)

Three of the head-shapes are being placed, one inside the other, so that the sectioning slits are "staggered." In this plant, the helmet liner pre-forms travel along a conveyor above the operators, to the molds. On the table, opposite the operators, two newly molded liners with excess material can be seen. (Photo courtesy of the U.S. Army Quartermaster Museum)

The operator is about to place a pre-formed helmet in the molding press where it will be given two minutes to cure at 220 degrees, under a pressure of 150 tons. Above the operator are the two conveyor lines, one of pre-formed liners and the other of newly molded liners. (Photo courtesy of the U.S. Army Quartermaster Museum)

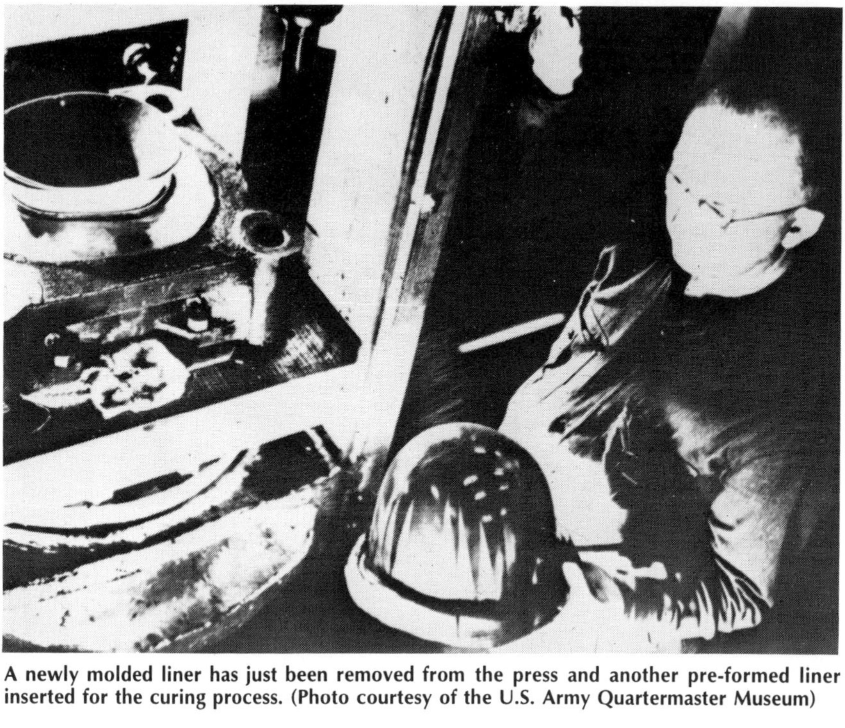

A newly molded liner has just been removed from the press and another pre-formed liner inserted for the curing process. (Photo courtesy of the U.S. Army Quartermaster Museum)

This liner has gone through the punch press to remove "flash" or excess material around the brim. (Photo courtesy of the U.S. Army Quartermaster Museum)

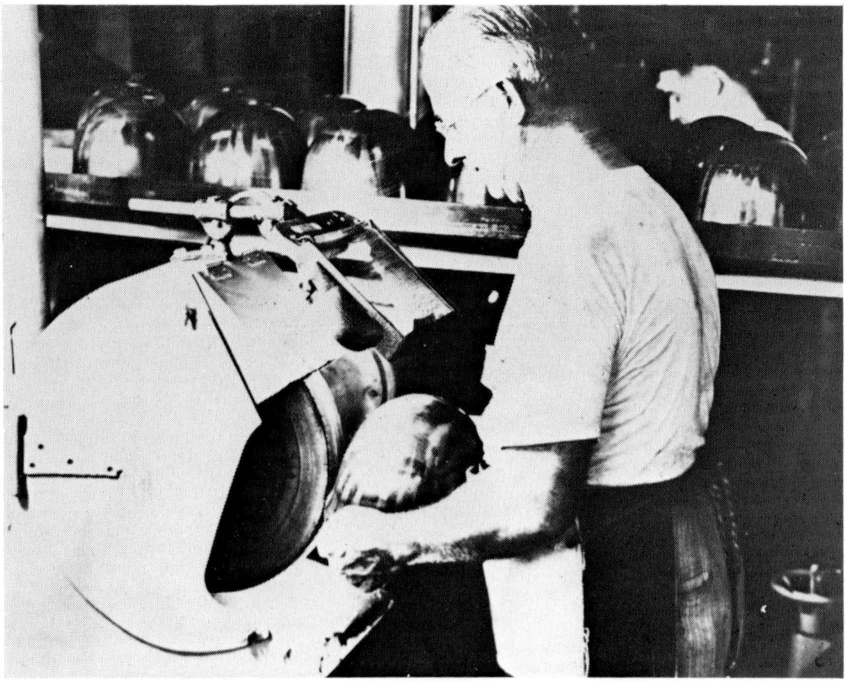

The raw edge of the liner is being burnished to bring about a heat seal. Several liners, one capping the other, could be burnished at one time. (Photo courtesy of the U.S. Army Quartermaster Museum)

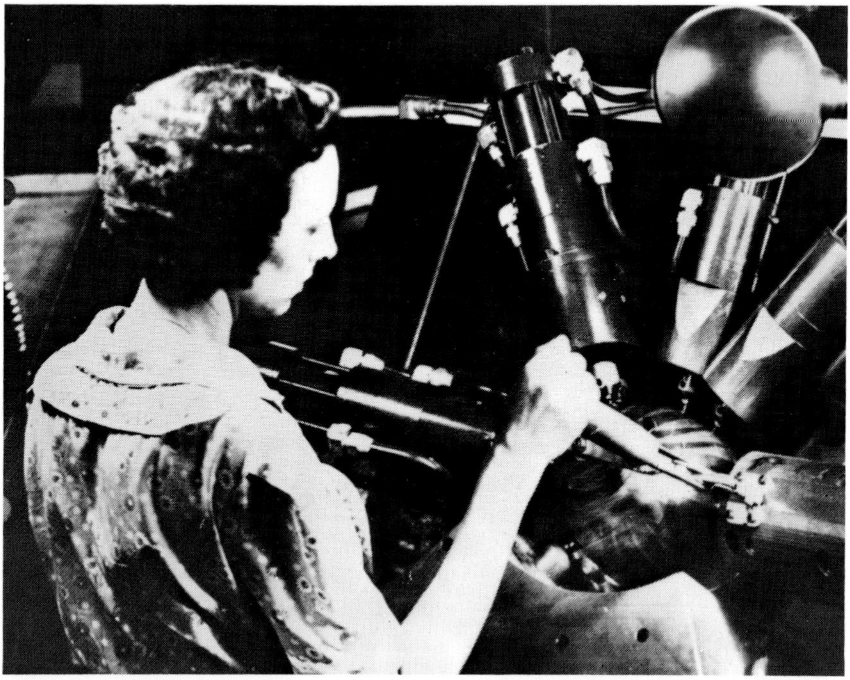

This liner is in the piercing press which punches all 11 rivet holes (standard M1 liner) in one operation. These holes admit the rivets and eyelet. (Photo courtesy of the U.S. Army Quartermaster Museum)

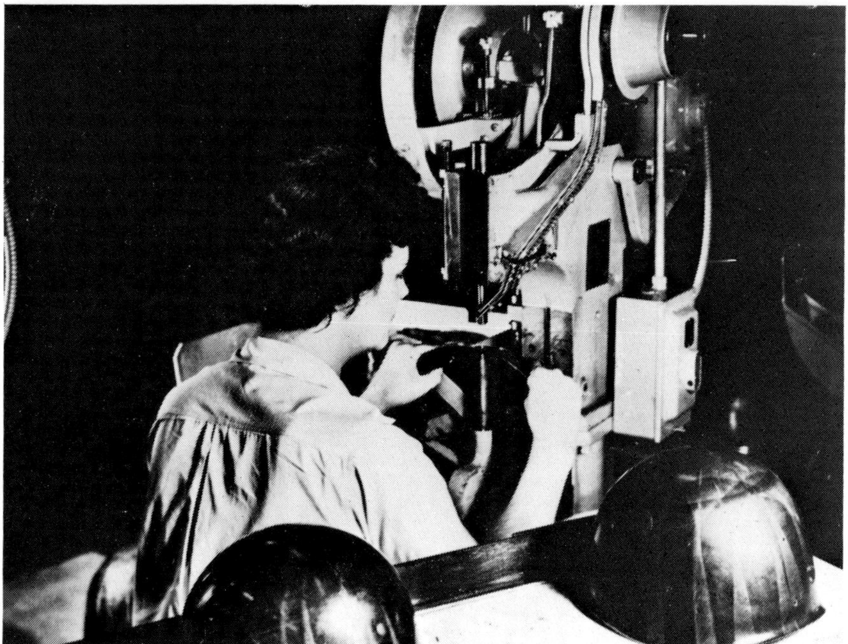

One of the riveting operations. Separate machines along the line handled the eyelet, the cap tuck and garter stud to carry the leather chinstrap, and other rivets fastening the suspension and neckstrap. (Photo courtesy of the U.S. Army Quartermaster Museum)

A paint-spray machine handled by one operator coats a liner set on a revolving tray. Some contractors make the coating operation completely automatic with the liners passing on a conveyor belt between two spray guns. (Photo courtesy of the U.S. Army Quartermaster Museum)

The liners, on a conveyor belt, are moving slowly though an infra-red oven. It took about two minutes to complete this operation which baked the paint onto the helmet liner surface. (Photo courtesy of the U.S. Army Quartermaster Museum)

Final inspection of the helmet liner following painting operations. (Photo courtesy of the U.S. Army Quartermaster Museum)

Liners being packed for domestic shipment. Complete with suspensions and chinstraps, they are packed in nested stacks of five with paper separators between two liners. Thirty liners were shipped in a box. (Photo courtesy of the U.S. Army Quartermaster Museum)

above the brim. Fitted with a metal grommet, this eyelet could be used by a soldier to affix rank or unit devices.

The liners then entered an assembly line where the chinstrap clip or stud, and the rivets needed to support the neck strap, were installed.

Painting was the final step. This was done by spraying an olive-drab finish, mixed with resin for texture, onto the shell. The shells were heated in infra-red oven for 15 minutes to help the paint dry.

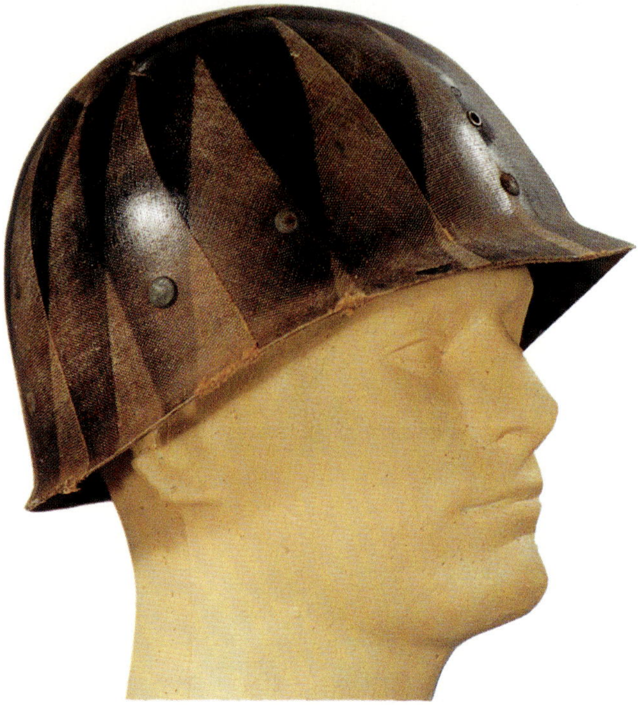

This unpainted high-pressure M1 helmet liner shows the intricate patterns of the duckcloth.

No Mark **Hawley Products Company**, St. Charles, llinois	HR **Hood Rubber Company, Inc.** Watertown, Massachusetts	SC **St. Clair Rubber Company** Marysville, Michigan	CAPAC **Capac Manufacturing Company** Capac, Michigan	F **Firestone Tire and Rubber Company** Akron, Ohio
IMP **International Molded Plastics,Inc.** Watertown, Massachusetts	S **Seaman Paper Company** Chicago, Illinois	W **Westinghouse Electric Company (Micarta Division)** Pittsburgh, Pennsylvania	**Mine Safety Appliance Company** Pittsburgh, Pennsylvania	**Inland Manufacturing Company** Dayton, Ohio

The Various maker marks of the eight companies producing M1 helmet liners during World War II.

In late 1942, 854, 225 helmet liners were painted in a camouflaged pattern. The helmet liners were envisioned for use by troops operating in jungle environments. The helmets were standard-issue M1 liners taken directly from stock. The OD paint was already applied and the camouflage pattern applied separately. A stencil was used for the pattern and the paint was applied using an airbrush system. Westinghouse accomplished the camouflage applications and used liners it manufactured. In 1944, the Quartermaster Corp. discontinued the camouflage liner program and ordered 300,000 liners repainted back to solid OD. (Courtesy of the Tom Gordon Collection)

Helmet liner painted medium blue for U.S. Navy use.

When the Army determined that liners made with the low-pressure method would no longer be acceptable for service many leftovers were sold as "toys" for children. This photo shows a St. Clair liner with the typical toy-style suspension.

Liner Components

Although manufacturing differences existed in the construction of the liner shells, the components which were installed into the helmets were basically the same for every manufacturer. These components included:

suspension
leather chinstrap
headband
neckstrap & neckband
shoestring adjustment
washers and rivets

The basic suspension of each liner consisted of a 1 1/8th-inch-wide band of silver/gray rayon webbing riveted around the perimeter of the interior. This band was fitted with 12 female snaps to accommodate the individually sized liner headband. Also riveted to the liner shell and sharing the same rivets as the suspension were three additional 1 1/8 inch-wide lengths of silver/gray rayon. The ends of these rayon loops were pulled to a position in the center of the helmet's crown and secured together with a shoestring-style cord. This resulted in a hammock type of suspension, which would support the helmet liner and steel shell.

At the rear of the liner shell another length of the 1 1/8th inch wide rayon material, approximately six inches long was riveted to the shell. This part of the suspension was fitted with four female snaps, two on each end, one over the other. These snaps were for the individually sized neckbands.

In June 1942, the material used in the liner suspensions was changed from the silver/gray rayon to khaki cotton webbing.[12] This change did not, however, affect the snap-in headbands as new helmets equipped with the khaki suspension continued to receive silver/gray rayon headbands. Apparently with a new headband design on the horizon, it was decided to use up existing stocks of the silver/gray snap-in headbands.[13]

Each helmet liner came equipped with a brown leather chinstrap to secure it firmly on the head. Besides filling the role as a liner for the steel shell, the helmet liner replaced a variety of hats for Army personnel. It was envisioned that while the entire helmet system would be worn in combat zones, in rear areas only the liner would be worn.

The chinstraps on the early Hawley liners and the earliest examples of plastic helmet were permanently attached to the helmet shell. The primary problem with the permanent chinstrap was that if it was damaged or broken, it was very difficult to repair or replace. During the middle of 1942, a garter stud-and-hook system became a standard helmet-liner component. This allowed the leather chinstrap to be easily removed from the liner.

Another change in the leather chinstrap addressed the buckle mechanism used to adjust it. The early, permanently attached chinstrap used a double-slit buckle. The leather was threaded through the buckle and was difficult

---

12 Collectors and researchers should note that manufacturers appear to have continued to use existing stocks of the silver/rayon after the June 1942 date.

13 It appears that snap-in headbands were not made in khaki.

to adjust. This adjustment buckle was replaced by a cam lever buckle which was much easier to operate.

In the field, the leather chinstrap was most often seen wrapped over the front visor of the steel helmet shell.

When early examples[14] of the M1 liner were shipped from the production centers to the field, all the components were installed, with the exception of the headband and the neckband. The reason these items were not included was because they were sized items which snapped into the internal suspension of the liner.

The headband for the helmet was snapped into place using 12 snaps positioned in double rows around the exterior of the headband. This item came in 13 sizes. The neckband, which was also a snap-in item, was made in three sizes.

The Army decided that to keep costs down, it would fit and issue the appropriate size headband and neckband when the helmet and liner were issued to the soldiers.

The need to stock 13 sized headbands was a nuisance the Army wanted to eliminate. Enter Garnett Skinner, the manager of Scholl Manufacturing

[14] *Liners made between 1941 and mid-1942.*

Straps, Suspension, Chin, Details, Liner, Helmet, M1 April 12, 1944

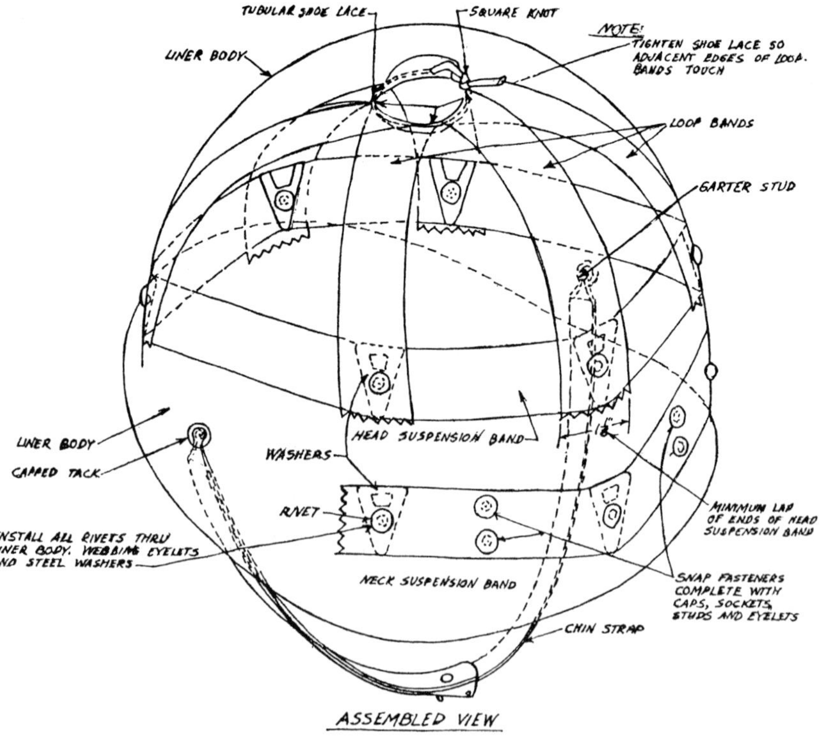

This specification diagram provides some excellent cutaway views of important parts of the World War II helmet liner. (Courtesy of the US Army Quartermaster Museum)

Company. He provided a sample of a one-size-fits-all headband that used a buckle-bar adjustment mechanism. The headband was made in one piece, like a belt. One end passed through the buckle-bar and made a circle which could be adjusted to fit any size. Skinner also developed a simple clip which could attach the headband to the suspension, thus eliminating the snaps on both the headband and suspension.[15]

These innovations, like the change from silver/gray rayon to khaki webbing, became standard in June, 1942.

Changes even affected the basic hardware used in the liners. Initially, the washers used to secure the internal suspension inside the shell were semi-rectangular and were uncomfortable. These were replaced by washers, developed by Westinghouse, which resembled the letter A.

---

[15] *Apparently Skinner got the idea from the method used to close a book of matches.*

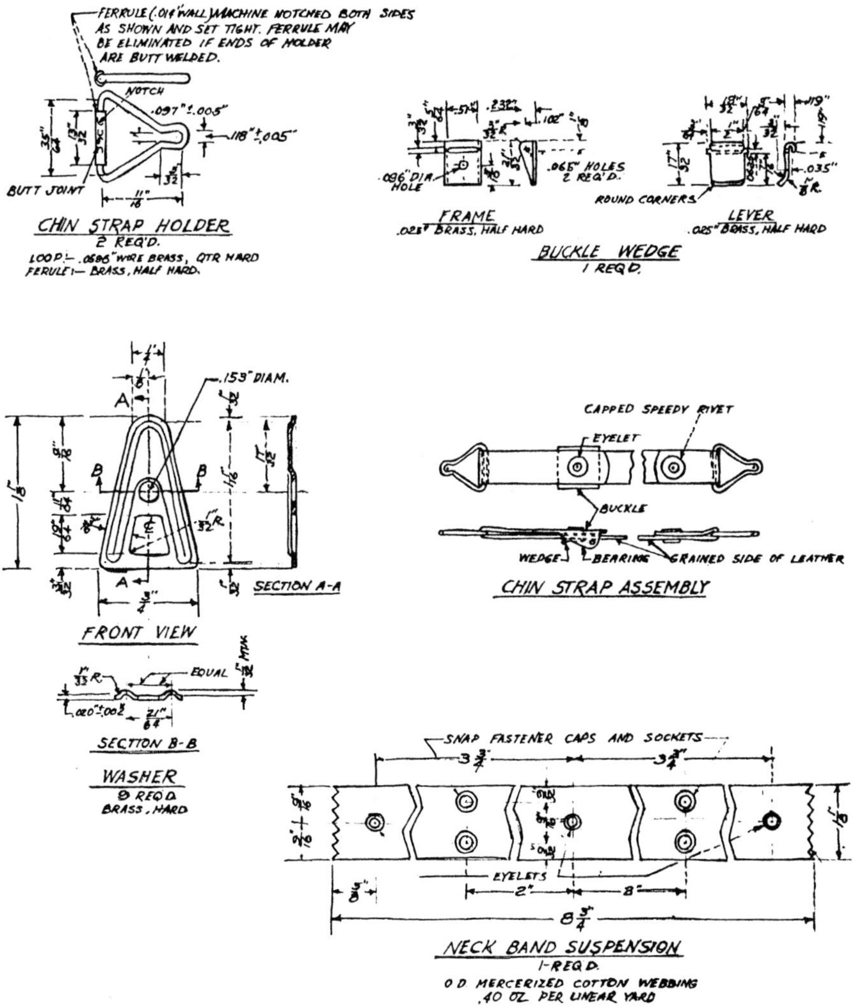

Leather M1 liner chinstraps. This type of connection replaced the early permanently attached clips. The straps were held in place by placing the garter clips around stud posts. The cam buckle allowed the strap to be adjusted. Early examples were painted OD but were later changed to a black corrosion resistant finish.

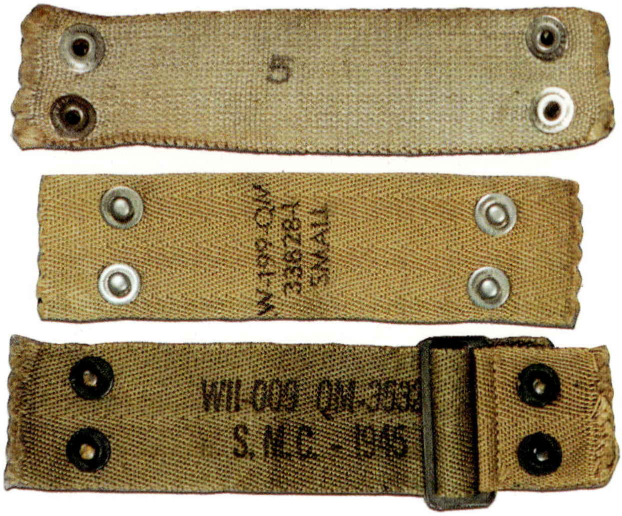

An assortment of neckbands. From top: Early silver/gray sized rayon, (note the number 5). Next, the khaki version made in three sizes: Small, medium, and large. Finally an adjustable neckband which eliminated the need for multiple sizes.

M1 liner headbands. These also evolved. At the top is the early silver/gray rayon snap-in headband made in 13 sizes. At the bottom, the late war, khaki, clip-in, adjustable headband.

The evolution of the WWII headband adjustment mechanism. Right: non-adjustable early sized example. Center: sliding clamp and bar buckle. Left: Late-war sliding buckle with clamp teeth. Note the clips used to hold the headbands onto the suspension.

A wide variance of colors can be encountered. The strange dark blue and brown vinyl headbands may have been used with St. Clair and Hood Rubber Liners.

Helmets for Parachute Units

As early as 1941, as the Army continued to develop its airborne divisions, it was apparent that leaping from a perfectly good airplane caused problems for personnel wearing the standard issue M1. A common problem was the steel shell separating from the helmet liner. The standard chinstrap also caused extreme discomfort to the trooper when jumping.

Keeping the steel shell on the soldier's head was a priority and an innovative solution was devised.

On a standard infantry M1 helmet, approximately one inch of the web chinstrap was threaded through the chinstrap loop, folded over, and sewn to the main body of the chinstrap. This fold secured the strap to the loop on the helmet shell.

To help hold the two parts of the airborne helmet together, the one-inch part of the strap threaded through the loop was extended to approximately three inches[16] and a male snap fastener was riveted to the end. At the one-inch mark, a row of stitches secured this short end to the main body. The remaining two-inch tab with the snap fastener could then be mated with the female part of the snap fastener just inside the helmet liner.

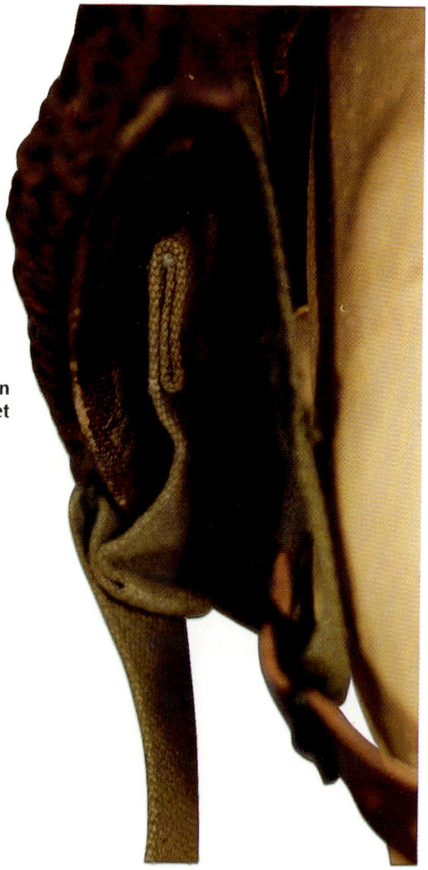

The additional tab on the M1C chinstrap can be seen snapped in position inside the helmet liner. (The Leffel Collection)

[16] *With the exception of the additional loop of webbing, the main chinstrap on the M2 was essentially the same as the M1.*

A M1C steel shell mated with an Inland-made paratroopers liner. The tab on the steel shell chinstrap is mated with the snap on the inside of the liner. (Author's collection and The Leffel Collection)

Another design change was to the helmet's chinstrap loop. Due to a modification on the airborne helmet's liner, it was determined that the chinstrap on the airborne helmet would not be used during parachute operations. During jumps, the standard M1 chinstrap would be pulled to the rear with the two ends clipped together above the helmet's rear rim.

General Frederick, 1st Special Service Force, wearing a M2 paratrooper helmet. Observe how the steel pot's chinstrap is worn positioned over the rear of the helmet shell. This was the specific purpose of the extra large "D" shaped chinstrap loop welded on the shell. (U.S. Army photo)

To make storing the chinstraps during jumps an easier operation, the shape of the chinstrap loops on the steel helmet shell was changed. Rather than standard square shaped chinstrap loops, the parachutist's helmet had half-round or D ring chinstrap loops. This modification resulted in the designation of the parachutists helmet as the M2 helmet.

Originally, the first airborne helmets were designated as M2 helmets. They used the larger "D" ring style of welded-on chinstrap loop. The liners for these M2 helmets were standard Hawley fiber liners which were modified by the addition of "A" frame webbing for a chinstrap assembly. (Courtesy of Ed Hicks)

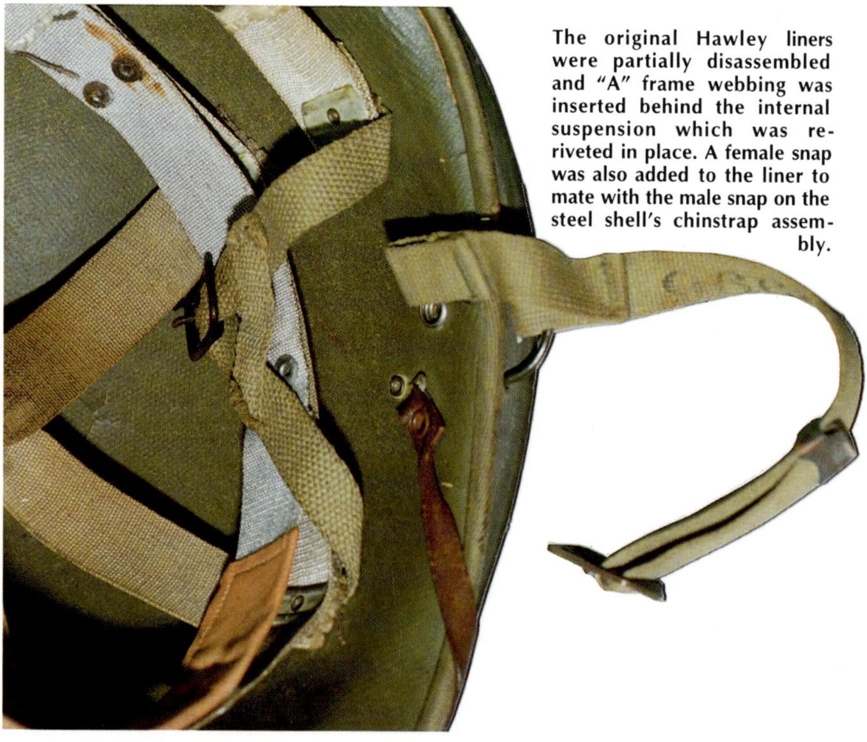

The original Hawley liners were partially disassembled and "A" frame webbing was inserted behind the internal suspension which was re-riveted in place. A female snap was also added to the liner to mate with the male snap on the steel shell's chinstrap assembly.

This new design was very short lived. Shortly after the production of the M2, another change was enacted.

Although the exact date is unknown, the Army determined that half round chinstrap loop was not required to secure the chinstrap at the rear of the helmet. (Possibly, M2 helmets, like early M1's were suffering damage to the chinstrap loops when dropped or mishandled.) The 1943 transition from the fixed, welded chinstrap loops to the flexible, hinge style loops probably terminated production of the M2.

To replace the M2, the Army turned to the original M1 helmet. Now with flexible hinge-style chinstrap loops, the airborne-specific chinstrap with tab and fastener was added and the helmet was designated the M1C. The M1C remained the airborne helmet for the remainder of the World War II.

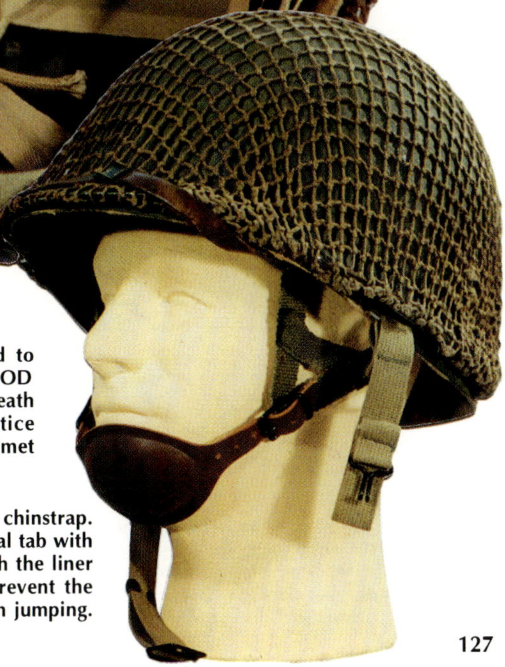

Above: A Capac-manufactured liner converted to M1C configuration by Westinghouse. Note the OD number 7 "A" frame webbing is secured beneath the original suspension webbing. This practice would be eliminated in post-war airborne helmet liners. (The Leffel Collection)

The M1C airborne helmet with early leather chinstrap. The helmet's original chinstrap had an additional tab with a male snap end. This tab was threaded beneath the liner and snapped in place. The purpose was to prevent the steel shell from separating from the liner when jumping. (The Leffel Collection)

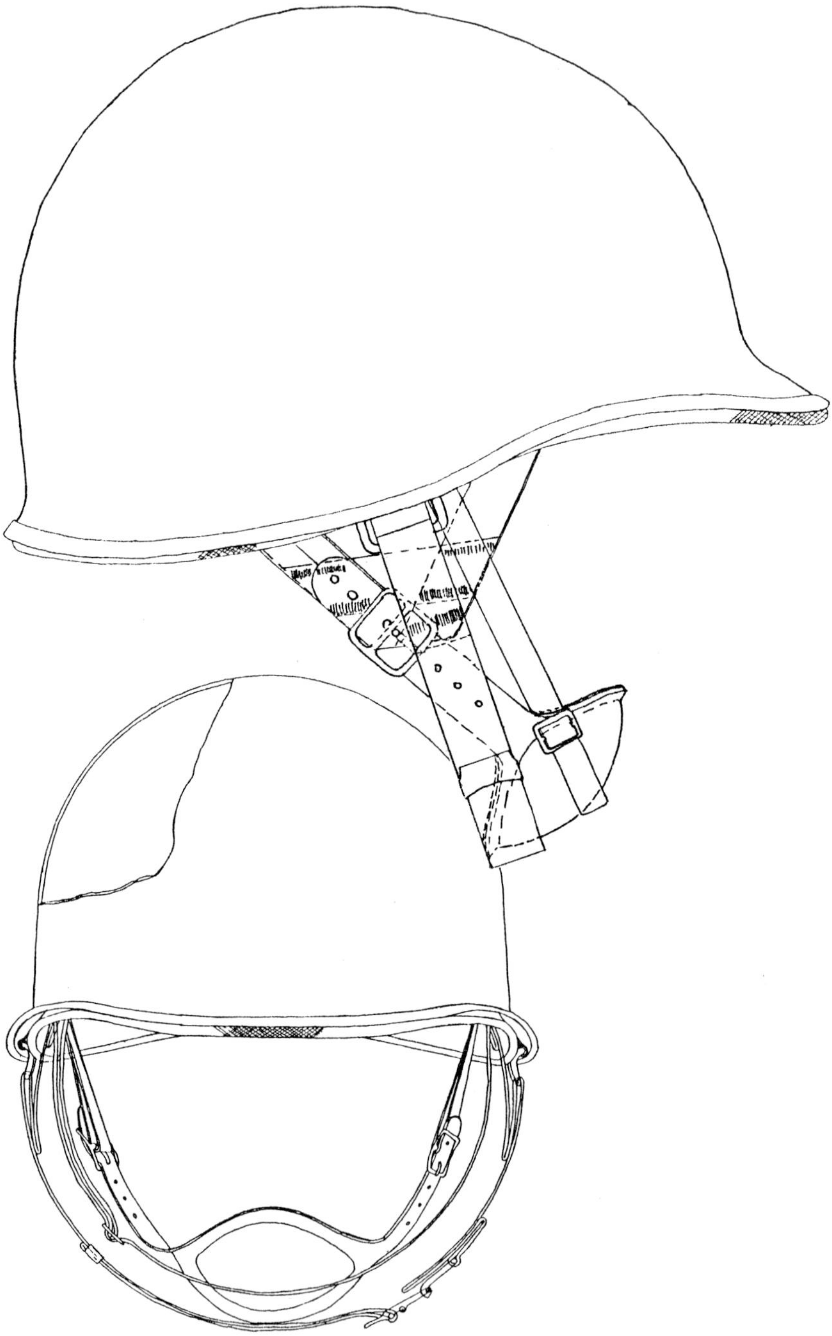

While we know this helmet as the M1C, the specification clearly has it labeled as a M2. Notice the chinstrap loops are the square hinged style and not the welded-on half-moon-shaped types of the M2. Both views provide an excellent look at how the chinstrap assembly appears on a paratrooper configured helmet. Notice from the front view how the chinstrap tabs loop inside the helmet shell to snap into the liner. (Courtesy of the U.S. Army Quartermaster Museum)

A M1C helmet with Hawley fiber liner. Note the first type of leather chin cup which was quite flat and did not fit the contours of the chin well.

An unknown soldier assigned to the 1st Special Service Force stationed in Alaska stands ready for inspection. He sports a M2 helmet with a plastic liner. His helmet shell is covered with the Canadian Style number II helmet net. (U.S. Army photo)

The Airborne Helmet Liner

Having solved the problem of the steel shell separating from the liner by riveting two female snaps to the inside of the liner (many soldiers reinforced this by putting tape on the front visor of the helmet and liner), the next problem was making the airborne helmet more comfortable and sturdy.

The comfort problem was due primarily to the helmet's instability during the jump. The standard M1 chinstrap was not robust enough for this task, so an alternative was designed.

Rather than make extreme modifications to the helmet's steel shell, an innovative solution using the helmet liner was developed. Instead of two chinstrap anchor points, four points, two on each side would greatly improve the helmet's stability.

The enhancement involved building a new chinstrap system directly into the liner. On each side of the liner, an inverted "A" frame constructed of khaki webbing was installed. At the end of each "A," a buckle with a tongue was positioned. A one-piece adjustable leather strap with chin cup was buckled into position on each side of the helmet. The wearer could loosen or tighten it.

An internal view of the Hawley liner reconfigured for airborne operations. (Courtesy of Ed Hicks)

The earliest parachutists' liner made for the M2 helmets was a modification of the original M1 Hawley liner. Approximately 43,000 Hawley's had the khaki inverted "A" frame chinstrap support webbing installed.[17]

Other early modifications included 75,000 plastic M1 liners made by Inland.

These early parachutists' liners can be identified by the construction method. The liners were partially disassembled and the supports for the front of the "A" frame shared suspension rivets. The rear of the "A" frame shared rivets which supported the neckband.

One of 75,000 M1C liners made by Inland. The M1C liner incorporated its own chinstrap webbing in an "A" frame configuration. The "A" frame webbing was inserted beneath the original suspension rivets on WWII examples.

Later in the war, parachutist liners would be made by Westinghouse. While the suspensions in these liners used khaki webbing, the "A" frame webbing exists in khaki and olive drab. The primary difference between these late war liners and the early Hawley's and Inland's is the majority have the inverted "A" frame webbing riveted to the helmet liner shell using the liner's original rivet points. Late war models made by Westinghouse have 17 rivets in the liner body rather than 13 rivets.

In late 1944, the leather chinstrap and cup used in the airborne helmet were replaced by a cotton, two-part, khaki chin cup equipped with four eyelets on each side for size adjustment.

---

[17] Finding an original M2 helmet with a Hawley manufactured liner converted for airborne use is extremely rare.

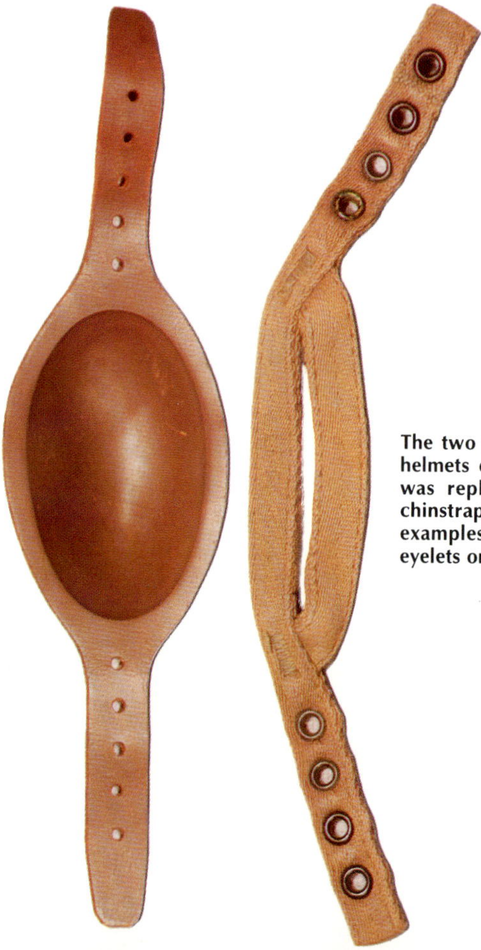

The two primary chinstraps used on airborne helmets during WWII. The leather chin cup was replaced in late 1944 by the cotton chinstrap primarily because of costs. WWII examples of the cotton chinstrap used four eyelets on each side.

An example of the WWII canvas chinstrap for airborne use.

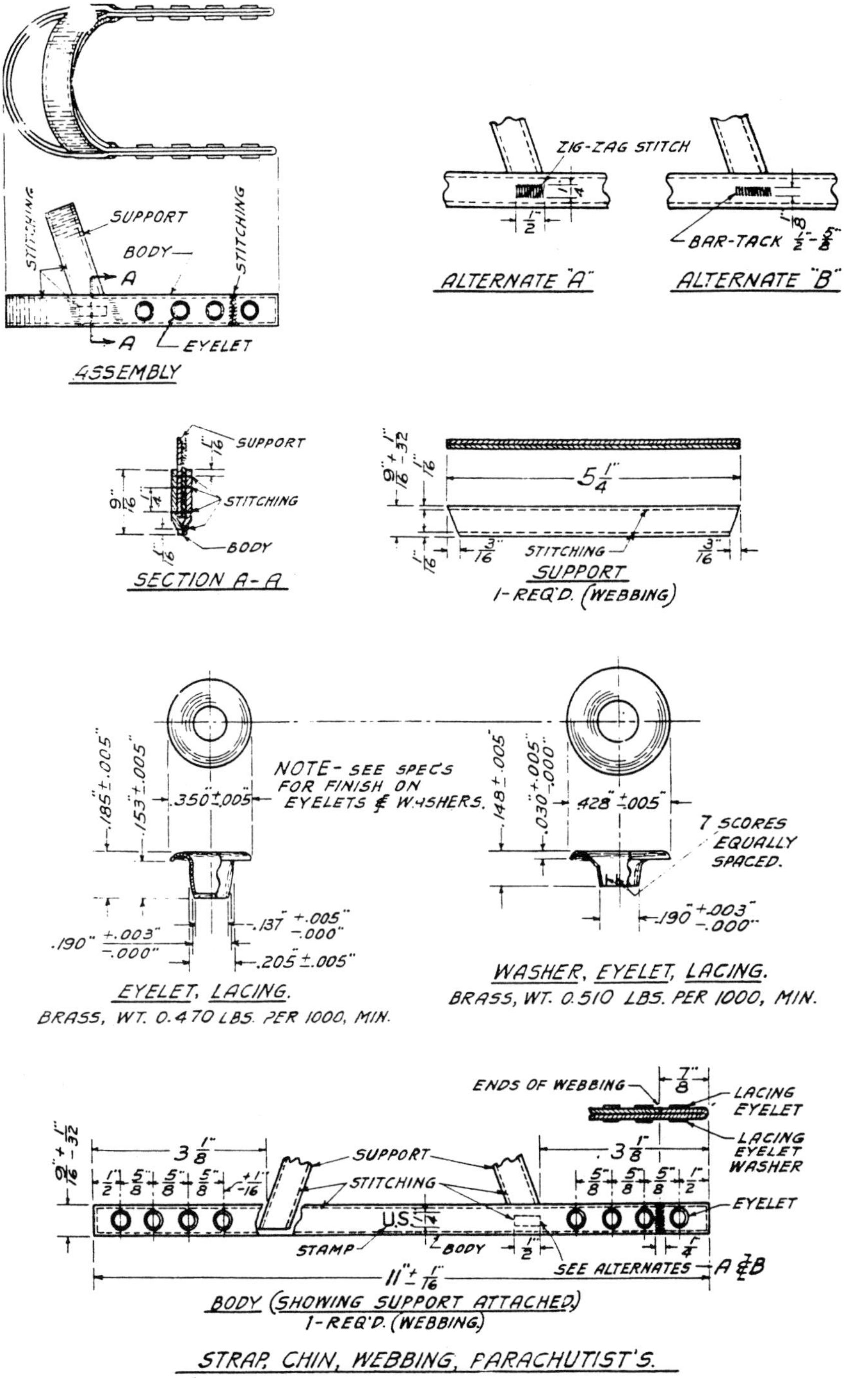

This 1944 specification details the construction of the webbed chinstrap which replaced the leather strap used on M1C helmets. Notice the chinstrap used four eyelets and was marked "US". (Courtesy of the US Army Quartermaster Museum)

Camouflage and Nets

Even with its cork textured, dark olive drab paint, the M1 helmet had a tendency to glare when wet. This is less than desirable for combat soldiers. Several solutions were considered to solve this problem.

First, the Army tried an alternative method for painting the helmets. Over the course of a year, approximately 850,000 M1 helmets and liners were painted in a standard camouflage pattern by Westinghouse. The paint was sprayed onto the helmets, which were covered by a stencil.

While Quartermaster personnel tested the camouflage painted helmets, combat soldiers in England discovered an effective method to prevent glare by borrowing an idea used by British soldiers; the helmet net.

Thousands of GI's procured netting material, placing it over their helmet. The net was held in place by sandwiching the excess material between the steel shell and the liner. The British nets were in assorted sizes but the most common consisted of an olive drab or khaki colored string with a 3/4 inch weave.

These troops, shown during the landings at Normandy on June 6, 1944, all wear the British/Canadian Type II helmet net. (US Army via George A. Petersen)

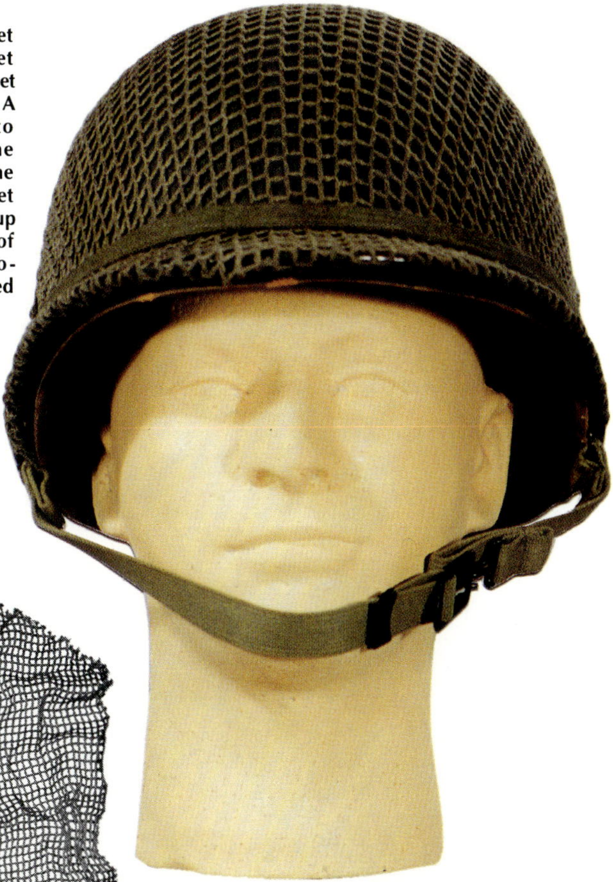

This is the US Army "Type I" helmet net with band. After testing, this net became the only official issue net procured stateside during WWII. A series of tests was conducted to determine which net provided the greatest camouflage capabilities for the troops. They determined that this net was the strongest wearing and broke up the glare of the helmet well. Some of these nets were issued without neoprene bands which were not considered as effective.

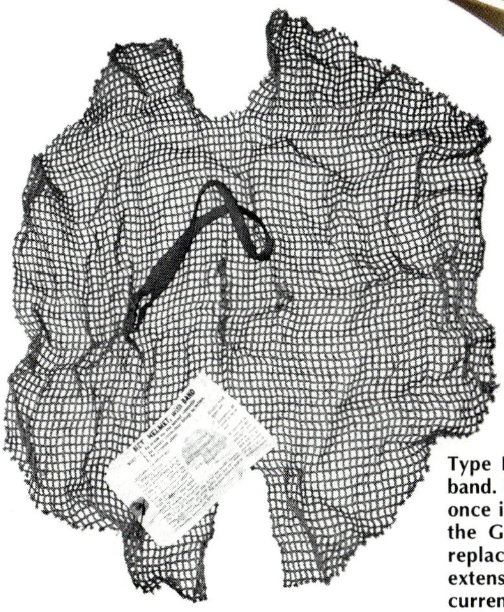

Type I helmet net as issued with a non-elastic band. Since the band was difficult to put on, and once it became worn or wet was not practical, the GI often removed it. After WWII it was replaced with an elastic band and was worn extensively in Vietnam. An elastic band is also currently worn on the Kevlar helmet. (Courtesy of Col. Barry Smith)

Just after the Allied invasion of Europe, the Army began testing an American helmet net. Constructed of cotton and manufactured using the newly approved Army color called olive drab #7, (a dark OD green), the one-size-fits-all net was strong and easy to use.

Like the British nets, the U.S. net was held in place by placing the excess material between the steel shell and the liner. It also came with a neoprene elastic headband to help secure it to the helmet and to hold foliage. The weave of the U.S.-made helmet net with band was approximately 1/4 inch. This netting was shipped to Europe just before the start of the Battle of the Bulge.

Known to the Army testers as the "Type II", this netting was acquired from the Canadian Army. Examples examined revealed the nets were too small to properly fit the U.S. helmet. The helmet was also a two-toned color and the loose weave did not reduce the glare as well as the Type 1 net. While not officially adopted, this net saw widespread use by U.S. soldiers during and after WWII.

This netting is often seen in WWII-period photos and films. The netting was originally used for camouflaging vehicles, and was available in great quantities in the US and in Europe. Consequently, there are many variations in color and weave.

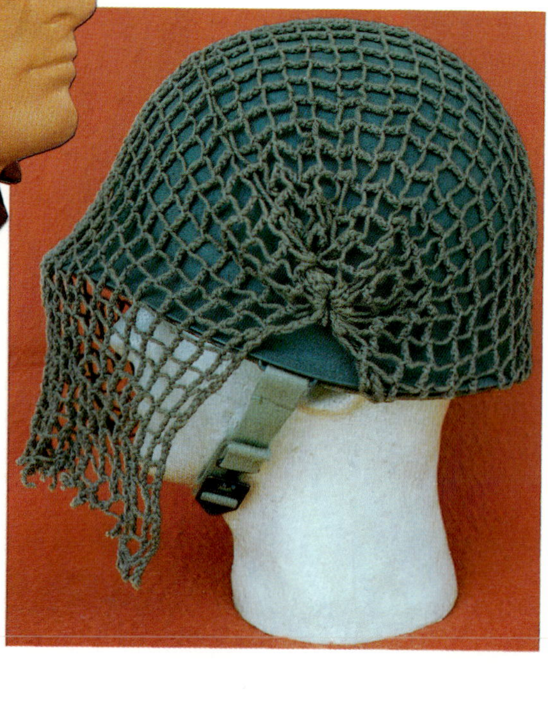

A rare example of a British-made commando net with face veil in the lowered position. (Courtesy of Bill Rizzutti)

COMMANDO H. NET.

A—Rosettes or gathers are approx. 1" behind chin strap and are under brim.

B—End of drawstring is under brim at back of helmet and drawn tight.

C—Drawstring in correct place in front of helmet.

SEE BACK FOR DIRECTIONS

DIRECTIONS FOR FITTING

1. Place helmet with peak outwards.
2. Place net over helmet with laced edge at back.
3. See that front cord (C) is in correct position (see picture). Draw strings tight at back and tie in channel just above back rim.
4. Force net at back of hat over rim as far forward as the rosettes or gathers (A). Final position on the Hat refer to illustration on back of this tag.

lo. 2928

ATE

ERATOR

NTITY

This is the instruction tag which accompanied the commando helmet net. The tag clearly displays a U.S. M1 helmet. The tag bears the famous British "Broad Arrow" markings. (Courtesy of Bill Rizzutti)

These 17th Airborne Division soldiers are wearing M2 helmets with an unknown type of helmet netting. (U.S. Army photo provided by Roger Bender)

With the net considered the fix to the glare problem, the Army soon canceled its experiment with camouflage paint jobs and Westinghouse terminated the project. Some 300,000 camouflage helmets and liners yet to be shipped to the Pacific Theater were repainted olive drab.

While the Army sought solutions for its soldiers, the procurement branch of the Marine Corps was also searching for methods to help its personnel blend into the Pacific jungles.

The Corps' answer to camouflage for the M1 helmet came in the form of a cover. During the Second World War, the Marines developed two patterns of reversible color helmet covers.

The covers were cotton in a herringbone twill pattern. The material was tough and would wear well. Like the helmet net, it would not add an unmanageable amount of weight when it was wet.

The cover was made in two identical shaped pieces, sewn together. The completed cover also included leaves or flaps which hung down below the main body of the cover. Like the helmet net, the camouflage cover was secured to the helmet by folding the leaves or flaps inside the steel shell and inserting the liner to hold them in place.

During World War II, the USMC developed three camouflage covers. Two were specifically designed as camouflage and the third had the dual purpose of camouflage and insect netting.

The early first pattern helmet cover, adopted in 1942, is made of a herringbone twill material and contains no button holes for twigs or foliage.

While the U.S. Army opted for nets to aid in camouflage, the United States Marine Corps used herringbone twill , cotton cloth covers. This is the first pattern cover and does not have foliage slits. It is reversible with a "green" side and a "brown" side.

It is a two-sided reversible cover which has a four-color pattern of greens and browns on one side while the reverse is three colors of tans and browns. The obvious intent would be for amphibious operations, the tan side could be worn when hitting the beach and the green side worn in jungle operations. History has shown the green side was worn much more often than the tan side.

The pattern which superseded this cover was nearly identical, except that 16 buttonhole slits were incorporated into the cover to accommodate natural camouflage items from the field. As with the early cover, this cover also was made in two pieces and sported a green and tan side.

This is an example of the second pattern USMC camouflage cover. The second pattern can be distinguished by 16 foliage slits in rows of two which circle the perimeter of the cover. This cover is also reversible and uses the same general colors. The eagle, globe and anchor insignia is probably a post-war addition to the cover.

Below: A comparison between the two different-colored sides of the USMC camouflaged helmet covers.

This second pattern stayed in USMC service into the 1960's.[18] Much of the camouflage material used in these helmet covers was produced by the textile firm of J.W. Valentine Company, Inc., New York.

The final camouflage helmet accessory used by USMC personnel was an insect net printed in a camouflage pattern. Often described incorrectly as a

---

[18] *Authors note: While USMC covers from the 1950's are often marked (normally with the 1953 date), I have never seen an authentic WWII era cover with date or maker's marks.*

A very common article of USMC helmet camouflage is the insect net. This item is often misrepresented as a "snipers veil" Of all the USMC camouflage covers, this is without a doubt the most common. That in itself indicates the sniper connection is incorrect.

"sniper's helmet cover"[19] this camouflage insect net was cotton, with a closely weaved mesh netting to protect the wearer from insects. Laces positioned at the bottom of the netting allowed the cover to be tied under the arms for increased protection. The camouflage pattern used on the insect net was similar to the green side pattern used on the USMC helmet covers. There appears to have been no tan side insect cover.

A close-up photo of a typical USMC eagle, globe, and anchor (EGA) insignia. These markings are usually post-war additions.

[19] *The sheer number of surviving unissued examples of this item makes it unlikely that they were made specifically for snipers. It was most likely an often-used, general issue item.*

Rarely if ever, was the EGA used on camouflage M1 helmet covers during WWII. This example, and most encountered by collectors, was probably added after WWII.

Below: A group of camouflaged Coast Guard cadets receive preparatory instructions before a simulated jungle attack. As future officers they trained with the troops they would one day ferry ashore in invasion barges. (US Coast Guard via George A. Petersen)

Constructed out of material similar to that used in Marine Corps shelter halves, the exact nomenclature or official description of this cover is not known. Often referred to as an experimental design, this camouflage is an experimental Army design and not that of the USMC! The cover has several very key differences from its two relatives. The cover's construction is far more complicated than the two standard covers previously mentioned in this book. Rather than rely on buttonhole slits built directly into the cover, reinforced patches with buttonhole slits were sewn on the front, rear, and both sides. Instead of a single seam down the middle, this helmet cover uses four sections of cloth sewn together. The inside of the cover contains a sand-colored, "brown side." (Courtesy of Major Jerry Lee, USAR, Retired)

This photo, dated February 1946, shows helmet covers similar to the one above being worn by US Army personnel. It was tested by the US Army but never adopted for general use. (US Army via George A. Petersen)

While the exterior of the cover had many differences in comparison to the more common USMC helmet covers, the method used to secure the cover to the helmet shell was even more different. Unlike the standard covers which used a system of leaves tucked between the liner and steel shell, this variant used two loops which wrapped around the steel shell chinstrap loops. The front and rear of the liner also had material which fitted the contours of the steel shell. All in all, little is known about these rare and unusual covers. (Courtesy of Major Jerry Lee, USAR, Retired)

This is another variant of the experimental Army helmet cover. The construction of the cover is identical to the other pattern but the camouflage is extremely different. It is very similar to the shades used on German Waffen-SS camouflage, which may have contributed to its demise. (Courtesy of the David B. Powers Collection)

The U.S. Navy also made use of the M1 helmets. All M1 helmets were originally painted OD but upon arrival at ships and ports many were painted various shades of dark blue and gray. It should be noted, however, that some specimens have been observed with what appears to be a factory painted finish.

The Navy Mark II Helmet

As with its sister services, the Navy made great use of the M1 helmet. From shore personnel to combat ships, the M1 helmet was the primary protective item of equipment.

When shipped from the assembly plants, the M1 helmets arrived in Navy hands in the olive-drab color scheme. But placed aboard combat vessels, most were repainted in battleship gray or dark sea blue.

One quandary was how to use large headphones sets for on-ship communications without altering the helmets. For a solution, the Navy contacted McCord in Detroit.

After considering the problem, McCord unveiled a new helmet for specialized shipboard use. Designed by Thomas J. Flynn and Albert Werheimer, the helmet, designated the MK 2, was an obvious cousin of the M1. A very deep, bowl-style helmet. It differed from the M1 in several respects:

First, the helmet's sides extended far lower than the M1, and since it was designed to accommodate headphones underneath it, was much wider.

Secondly, the chinstrap was vastly different. The MK 2 chinstrap was leather, very similar to the early leather straps used in airborne helmets. The MK 2 did not have chinstrap loops. Instead, two small holes were punched into each side of the helmet, just above the rim, about five inches apart. Into these holes went a small wire attached to a leather thong approximately six inches long. The leather chinstrap was then buckled onto the thong, which was held in place by the thin wires. The purpose of this chinstrap retention system was to permit strap release due to blast concussion.

Feb. 8, 1944. T. J. FLYNN ET AL 2,340,872

HELMET

Filed May 13, 1942 2 Sheets-Sheet 2

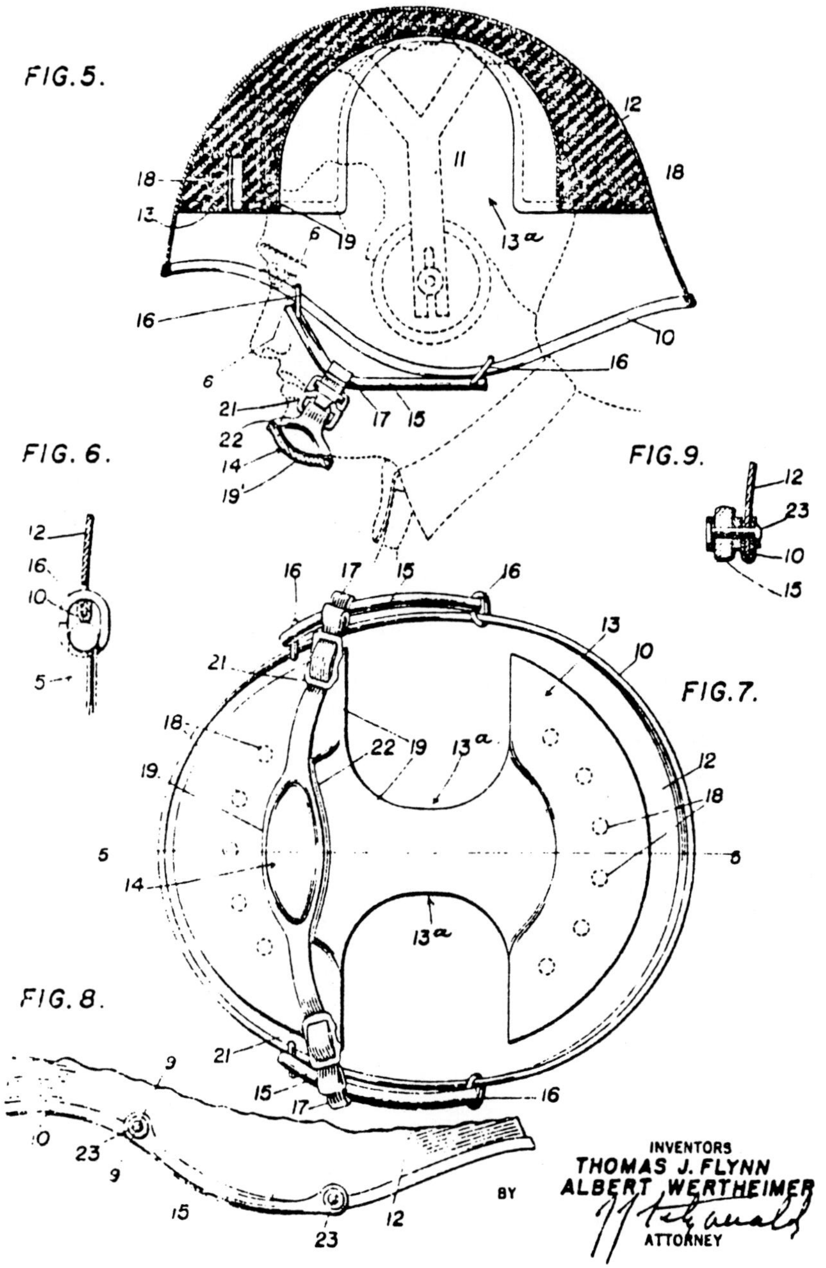

This drawing illustrates how headphones could be worn beneath the helmet. The striped area designates the internal padding and suspension. Notice how it is cut away where the headphones are positioned.

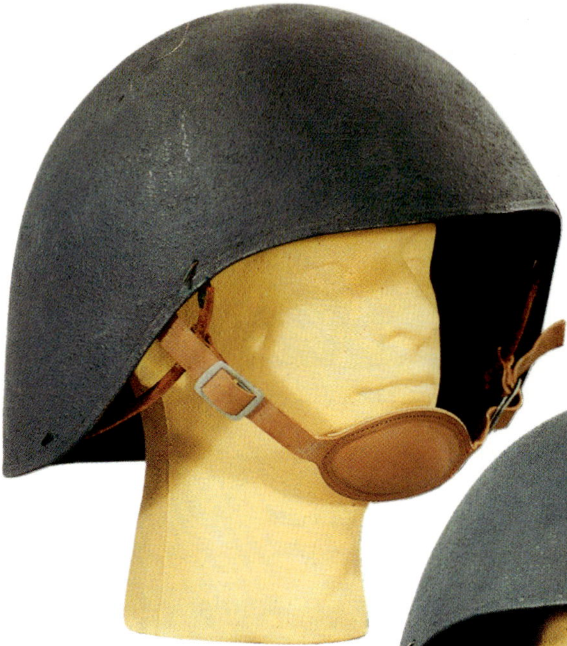

This is the MK 2 talker helmet developed in 1942. The helmet was created so that communications headgear could be worn with comfort under the helmet.

A front view of a US Navy MK 2 helmet. The MK 2 was an obvious variation on the M1.

The suspension system in the MK 2 is constructed from foam rubber covered with a water-resistant oilcloth. The suspension was glued in place and could not be removed from the helmet.

The final difference in the MK 2 was the liner. Unlike the M1, the MK 2's liner consisted of a grey colored cotton-covered, or waterproofed, shiny cloth-covered rubber padding. This padding was glued into the crown of the helmet. It was molded to fit the head and appeared to be made as a one-size-fits-all helmet. The padding supported the top, front and back of the head but did not cover the sides, so personnel could wear headphones while wearing the helmet.

McCord manufactured approximately 400,000 MK 2 helmets. The early MK 2 helmet left the factory painted OD but soon the helmet color changed to a dark bluish-gray color both inside and out. Inside the helmet the words U.S.N. and MK 2 are often found.

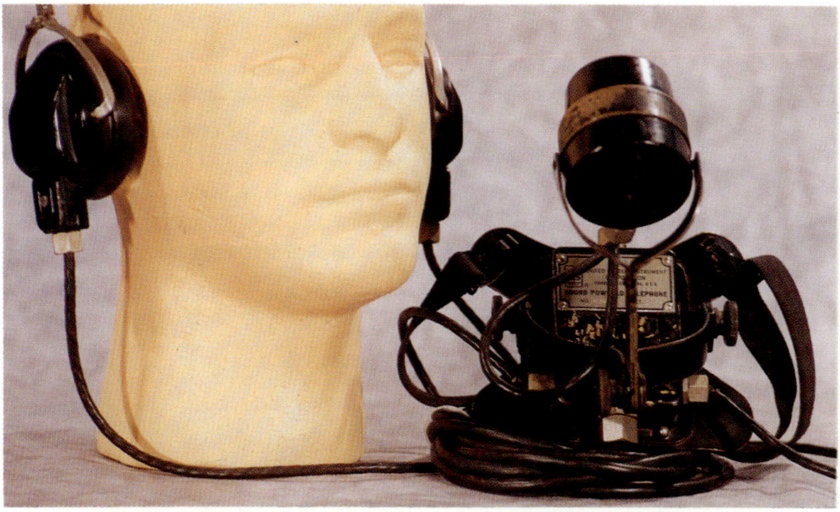

Headphones such as these were worn beneath the MK 2 Talker helmet. The microphone assembly rested on the chest allowing "hands free" operation of the mic.

The chinstrap retention system on the MK 2 was also very different from the M1. The chinstrap was buckled to a thong attached to the helmet shell. While quick release in the event of a nearby explosion was a potential purpose of this system, the most plausible reasoning for the system was to make storing the helmets on the ships more efficient. The chinstraps were removed and the shells could be stacked. The chinstraps were stored next to the helmets.

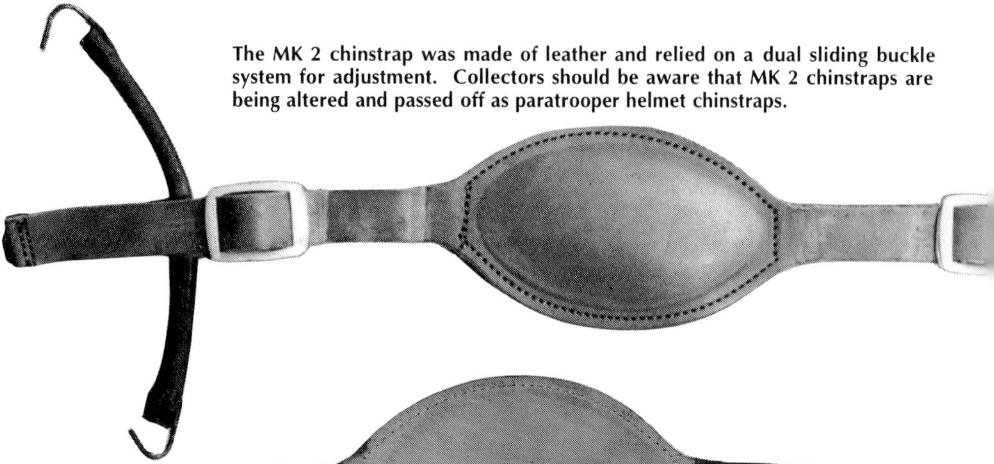

The MK 2 chinstrap was made of leather and relied on a dual sliding buckle system for adjustment. Collectors should be aware that MK 2 chinstraps are being altered and passed off as paratrooper helmet chinstraps.

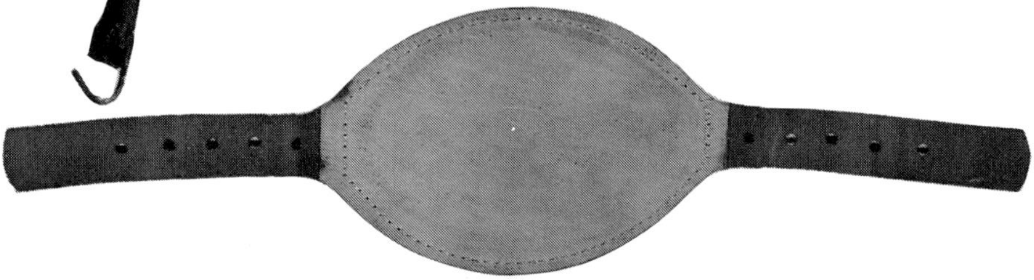

A MK 2 chinstrap altered to be a replica M1C chin cup for use by WWII reenactment groups.

Late War M1 Developments and Experimentals

The M1 helmet underwent constant change and revision during World War II.

Two changes instituted very late were the T-1 release and the T-1 chinstrap clip.

The T-1 release was an innovative but complex ball-and-latch mechanism used to secure the two ends of the chinstrap together. The original chinstrap clip consisted of a dual-hook catch on the short strap which locked into a pointed arrow prong on the long end. This system worked exceptionally well in keeping the chinstrap secure. In fact, it worked too well.

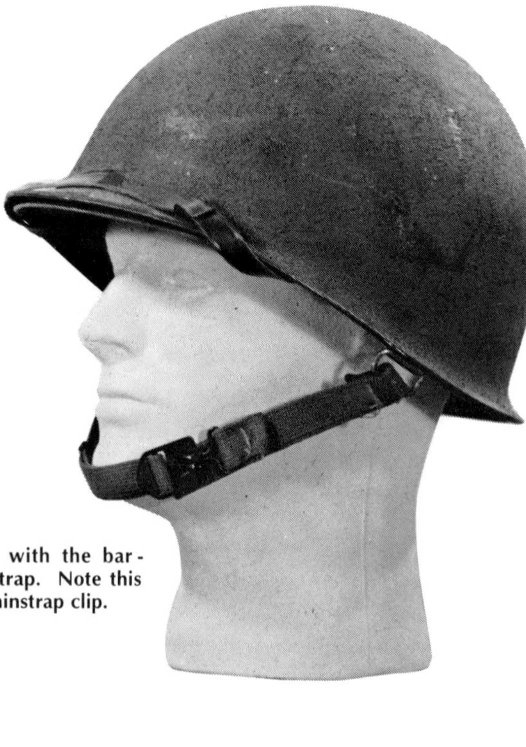

A late-war M1 helmet equipped with the bar-tacked olive drab Number 7 chinstrap. Note this helmet is not equipped with a T1 chinstrap clip.

The heavy khaki web chinstrap with arrow buckle and hook used on the M1 helmet.

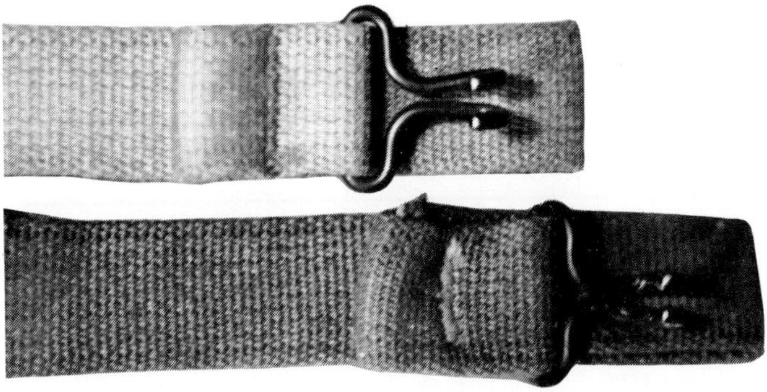

A comparison of the early khaki chinstrap webbing and the late-war olive drab Number 7 colored chinstraps.

Very disturbing reports from the combat zones indicated several soldiers had been killed or seriously wounded when high explosive artillery or tank fire detonated near them. The injury wasn't due to shrapnel or fire but the concussion. Apparently the concussion could be so great that the blast wave entered the helmet and could break the neck of a soldier if his chinstrap were buckled.

To alleviate that, the T-1 was constructed so that it would separate with only 15 pounds of pressure. Although approved for use in 1944, very few T-1 chinstrap releases ever made it to battlefields in Europe or the Pacific.

Replacement parts for M1 helmets were another irritant to the Quartermaster corps. Although most components were easily replaceable (like the entire helmet liner), the chinstrap sewn onto the helmet loops of the M1 was not. If the strap broke, a soldier had to locate a QM depot to have a new chinstrap sewn on the shell. A quick field fix was often more realistic during combat operations. One field repair suggestion for soldiers was to use a rivet as a temporary way to keep the strap in place.

While work was being done on the Army Air Force's M5 anti-flak helmet, the method used for attaching the chinstrap to that helmet was identified as a possible solution to the M1's chinstrap replacement problem.

The M5 helmet used a T-1 chinstrap clamp. Like the M1, the M5 used the same hinged chinstrap loops.

The T-1 chinstrap clamp was a fold-over clamp with small metal teeth which bit into the webbing of the chinstrap and was wrapped over the hinged helmet loop. It could be easily removed by using a knife, screwdriver, or a pair of pliers.

CHIN STRAP RELEASE, T1
(For Helmet, M1)

This release is designed to prevent injury to the wearer's neck by the chin strap due to forces on the helmet from concussion or sudden impact.

Separate spring hook (A) from ball tongue (B) by grasping each part firmly and pulling apart *(Figs. 2 and 3)*.

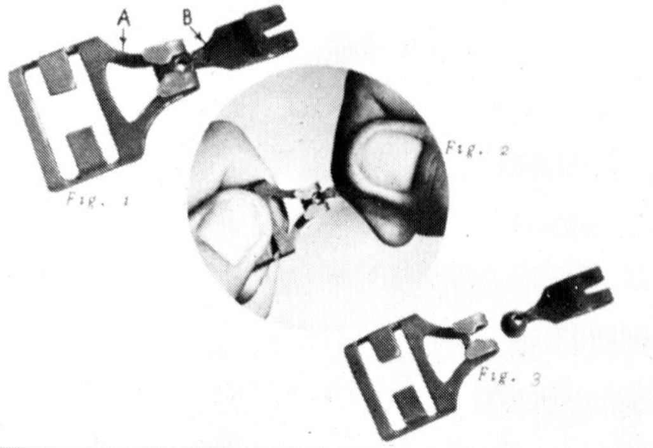

Fig. 1 Fig. 2 Fig. 3

Spread old hook prongs *(Fig. 4)*, work loose and then pull out of strap loop. If no tools are available, spread hook by using edge of helmet. Insert new hook in same position. Open ends only are inserted in loop, *(Figs. 5 and 6)*. Remove no other parts from strap.

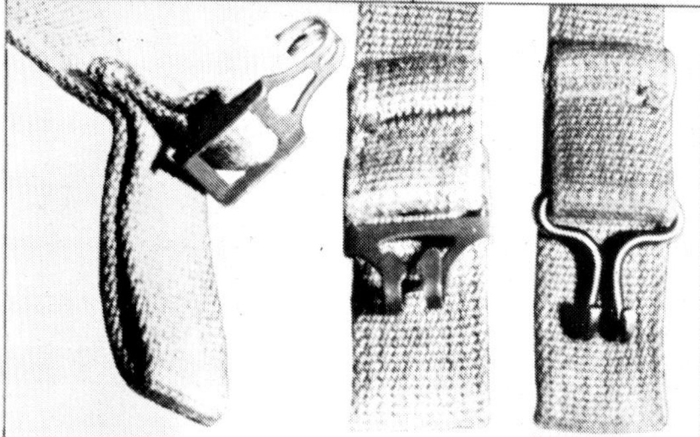
Fig. 6 Fig. 5 Fig. 4

Grasp ball tongue at the ball end, insert in buckle opening *(Fig. 7)*, and pull upward so that it hooks around cross bar [illegible] ball tongue should rest on outside of buckle *(Fig. 8)*.

Fig. 7

Fig. 8

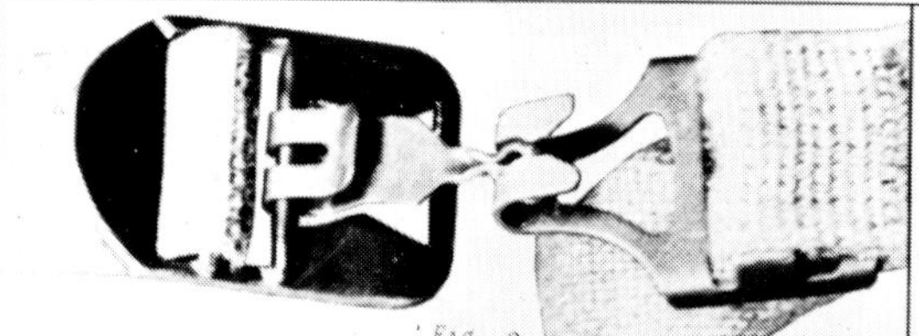
Fig. 9

To set chin strap in automatic release position, insert ball tongue into hook slot and pull tight *(Fig. 9)*.

After assembly of release to chin strap, engage and pull apart ten times to insure smooth operation.

WARNING — When helmet is being worn *DO NOT* fasten strap by hooking spring onto buckle. *CONCUSSION WILL NOT RELEASE CHIN STRAP IN THIS POSITION (Fig. 10).*

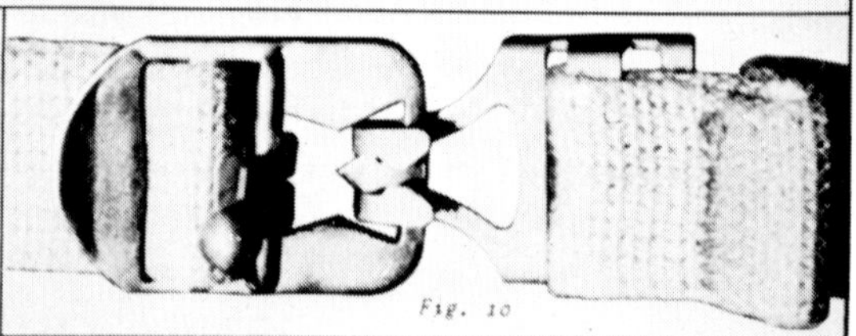
Fig. 10

QMB PROJECT T-376 EXHIBIT "C"

Like the T-1 chinstrap release, the T-1 chinstrap clamp was developed too late to assist the GI in World War II.

Besides protection, the American soldier, famous for improvisation, found an assortment of uses for his helmet. They became sinks, washbasins, shovels, laundry tubs, stools, and stoves.[21]

With 22 million M1 helmet shells and some 33 million liners made during the war, the M1 helmet, which cost approximately nine dollars, was credited with saving the lives of thousands of American service men and women in every theater of operation.

Since the War, dozens of countries have copied its design and to this day the same basic helmet is still in service with every component of our armed services.

A very interesting M1 helmet with a built-in suspension system. The exterior of this helmet conceals the fact that this steel shell has its own liner. A steel band has been welded into the shell to accommodate the web suspension. Since the helmet did not require a separate liner it was slightly smaller than a normal M1 helmet. This helmet retains all the characteristics of a M1 helmet pressed and constructed during World War II. The tag in the crown of the helmet provides the thickness of the steel before (.052) and after (.040) the draw. The helmet weighs three pounds, three ounces. It is unknown how many of these helmets were produced. (Courtesy of Bill Rizzutti).

◀ The T1 chinstrap release was created due to reports of soldiers being injured near blast concussions. The new ball-and-clip release was designed to separate with 15 pounds of force. (U.S. Army Photo, Courtesy of the Quartermaster Museum)

[20] *While many GI's apparently used the helmet as a pot to cook in or boil water, this greatly reduced the protective qualities of the helmet and rendered it unserviceable. Orders were issued and the practice was banned.*

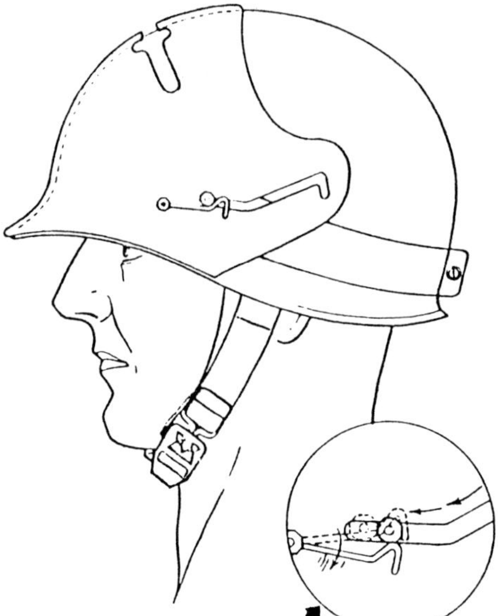

HOOK HAS SPRING ACTION TO RETAIN VISOR WHEN NOT IN USE.

SIDE VIEW - VISOR INACTIVE

This is a 1944 diagram of an experimental face shield for the M1 helmet. The shield itself was constructed from a M1 steel shell. Like experimental visored helmets of WWI, the shield could be worn in a raised or lowered position. A 1944 research project report confirms that at least one example was produced.

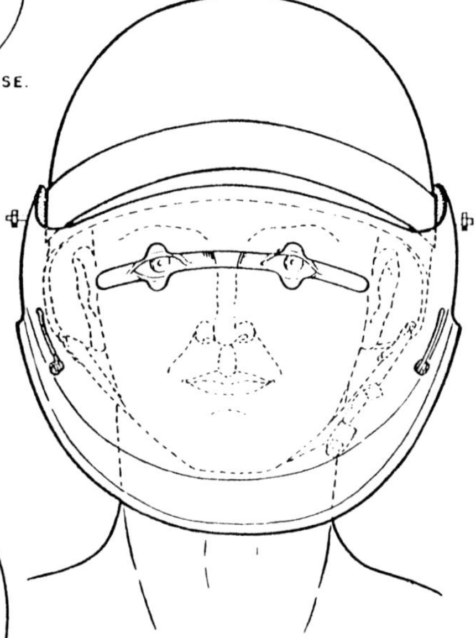

FRONT VIEW - VISOR ACTIVE

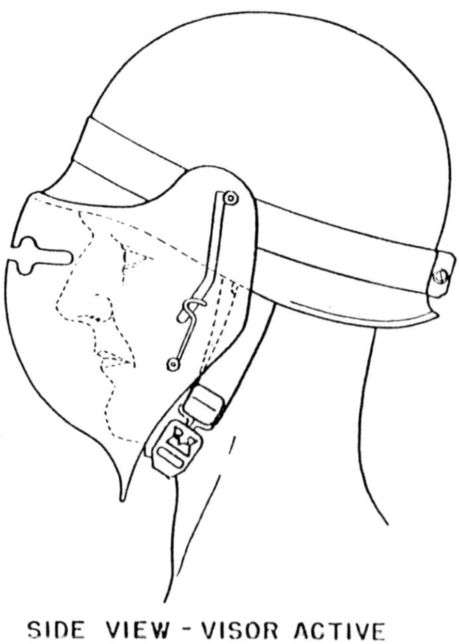

SIDE VIEW - VISOR ACTIVE

The T-14 was an experimental helmet based on the M1. It was designed for use by photographers and incorporated a shield which could be raised and lowered. Note the larger pivot point on the face shield at left. (Photo courtesy of David B. Powers)

Sgt. Marion Dutro, 165th Signal Photo Company, France 1944. In this rare photo, he is testing a variant of the experimental T-14 visored helmet. (Photo courtesy of "The GI Journal")

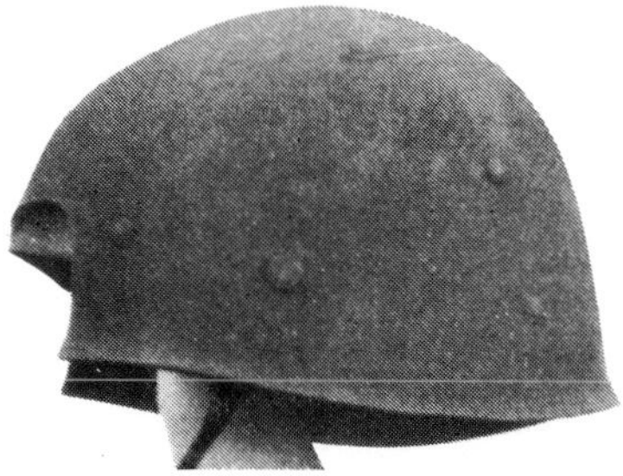

The liner of the T-14 was essentially a standard M1 helmet with the front brim removed.

Another modification resulted in testing and evaluation of a crash helmet for tankers and armored crew vehicles. The design was similar to a M1 liner with the brim removed and cut-outs added for earphones. (U.S. Army photo courtesy of the Quartermaster Museum)

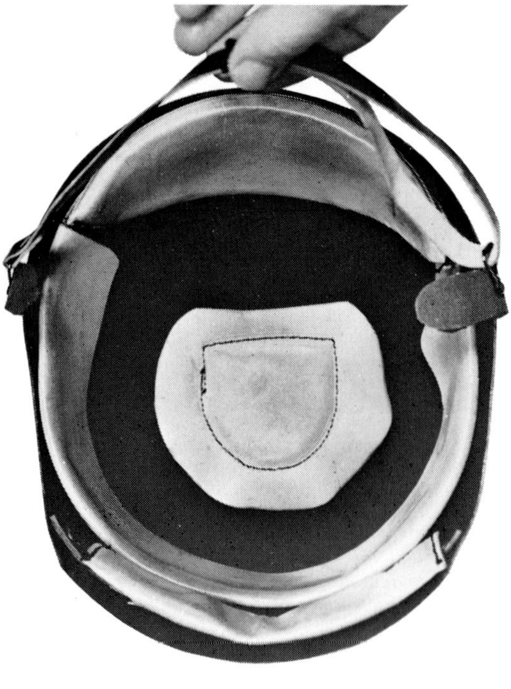

The experimental crash helmet used a very different system for suspension than the M1 liner. A series of leather covered pads was designed to cushion the impact if the wearer struck his head on one of the many components inside an armored vehicle, a very common occurrence. (U.S. Army photo courtesy of the Quartermaster Museum)

Another project to provide additional protection to tank and armored vehicle crews was the experimental model T-19E1. The idea behind the T-19E1 was to provide a helmet which could be used with an improved padded suspension and allow the wearer to used headphones. The front brim area was removed to aid in using weapons-sighting devices. While several were manufactured and tested, the helmet did not go into production. (U.S. Army photograph)

This is experimental helmet T-20E1 designed for tank crew members. Like the T-19E1 the liner system was cut away in a manner similar to the US Navy MKII helmet to allow the use of headphones. (Courtesy of Bill Rizzutti)

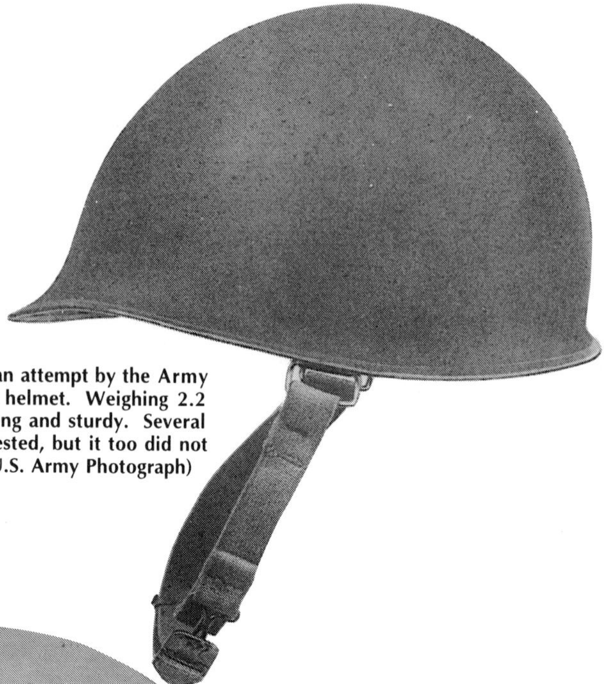

This helmet, the T-21, was an attempt by the Army to reduce the profile of the helmet. Weighing 2.2 pounds the helmet was strong and sturdy. Several examples were made and tested, but it too did not make it into production. (U.S. Army Photograph)

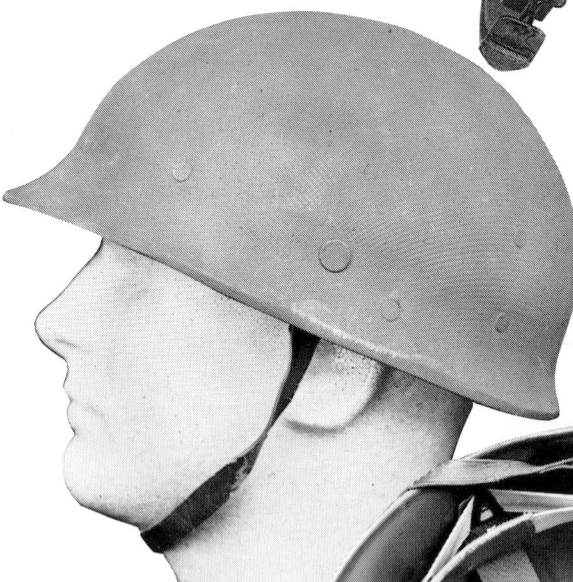

The exact purpose of this very shallow M1 style helmet liner is not known. It may have been the liner intended for use with the T21 helmet. (Courtesy of Bill Rizzutti)

Several types of experimental helmets designed for tank crews based on the USAAF M3 and M5 designs. The variations in these helmets primarily concern shape and method of manufacture. The helmets, designated T-10, T-12, T-13, and T-16 were tested and evaluated at Fort Knox, Kentucky in 1944 but they were rejected by the armored board.

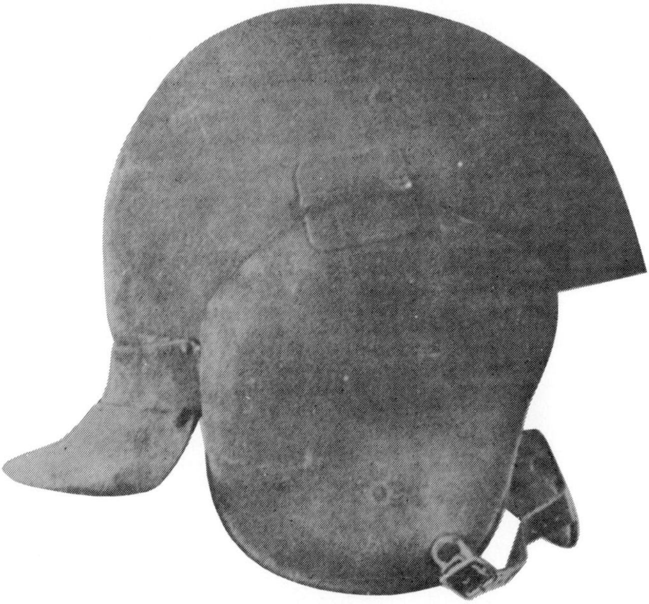

T-10

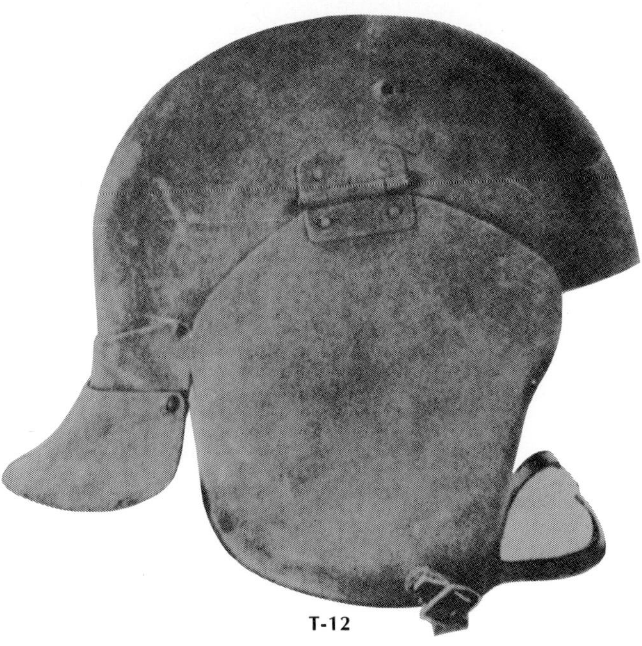

T-12

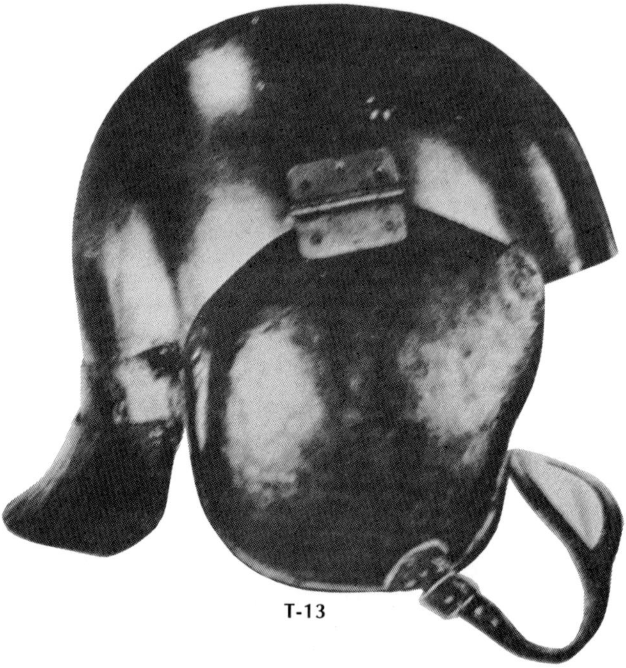

T-13

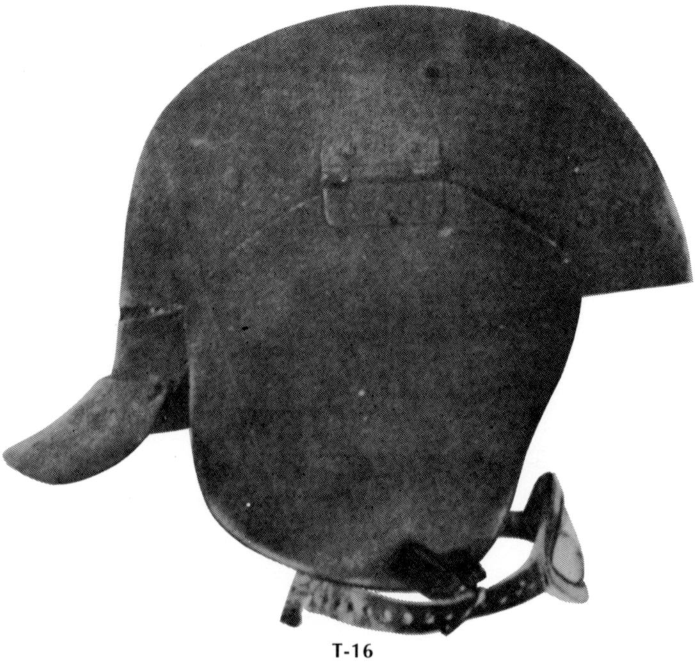

T-16

The sheer variety of M1 helmet configurations, chinstraps, liners, covers, nets and other accessories developed during World War II make this seemingly simple item of equipment quite a challenge to study and collect. Add in an element of individual embellishment, and the complexity is multiplied even more.

While the emphasis of this book is to showcase the variations and evolution of American combat helmets, liners and accessories, I would like to include a quick note about helmet markings. The wide variety of insignia and emblems painted on M1 helmets during World War II, coupled with the fact that standardization and official authorization was often non-existent, make this a challenging subject. A quick primer on field-marked M1 helmets reveals the most common markings are:

- A white vertical bar painted on the rear of the helmet identified the wearer as an officer and was seen primarily in the European Theater.
- A white horizontal bar painted on the rear of the helmet indicated the soldier was a non-commissioned officer, also seen predominately in Europe.
- Helmets with Red Cross insignia were worn by medics or doctors.
- Helmets with MP markings were worn by members of the Military Police.
- Individual divisional or unit insignia, such as the famous blue and white striped square of the U.S. 3rd Infantry Division.

The huge variety and lack of standardization make this subject ripe for study. With that in mind, the author would like to invite readers to submit photos for a potential follow-up book on the subject of field-marked/painted U.S. helmets from World War I to present. He will be seeking your photos of

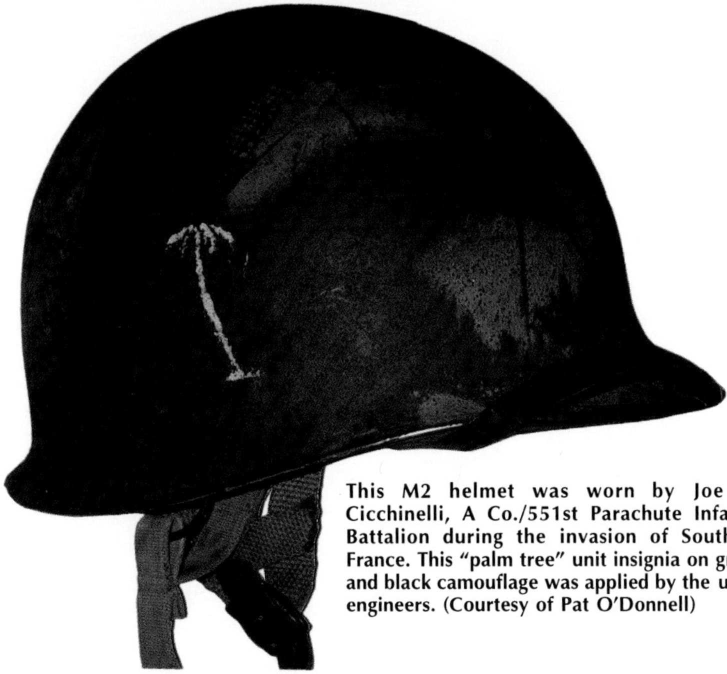

This M2 helmet was worn by Joe M. Cicchinelli, A Co./551st Parachute Infantry Battalion during the invasion of Southern France. This "palm tree" unit insignia on green and black camouflage was applied by the unit's engineers. (Courtesy of Pat O'Donnell)

Mac West

US Army Quartermaster Museum

Kimball S. Jeter

George A. Petersen

George A. Petersen

painted or marked U.S. helmets. Plus, if you have interesting period photos or information about unit markings we hope you will share those as well. To submit a photograph or information, send copies to: R. James Bender Publishing, P.O. Box 23456, San Jose, CA 95153-3456.

Please include a full, detailed description of the helmet plus any other details you might know about each helmet photo you submit.

R.J. Bender

Jim Phillips

Col. Barry Smith

Col. Barry Smith

George A. Petersen

Col. Barry Smith

CHAPTER FIVE

AIRCREW ANTI-FLAK HELMETS OF WWII

American soldiers on the ground and at sea were not the only ones in need of armor to improve their chances for survival. For U.S. airmen, combat over the skies of Germany and Japan was extremely hazardous. As if sub-freezing temperatures, oxygen starvation, and mechanical difficulties weren't arduous enough, enemy forces worked furiously to shoot down allied aircraft.

While enemy aircraft swarmed around bomber formations, another major threat was the effects of Triple-A, or Anti-Aircraft Artillery — Flak.

World War II aircrews often joked that "the flak was so thick we could walk on it." As unreasonable as it sounds, flyers never seem to be smiling or joking when repeating the phrase. Flak was a very serious threat.

American bomber forces were at great risk from flak and in 1942, head wounds accounted for nearly 40 percent of the fatal injuries. Obviously, lowering this statistic was a huge priority to the Army Air Force. No specialized protective headgear had yet been developed to protect combat aircrews, so the standard M1 helmet was pressed into service.

A major problem encountered was the aircrew's headsets. They were large and didn't fit well under the steel shell and liner of the M1. To remedy this, many men simply wore the steel shell on top of their leather flying helmet. Others refused to wear the "steel pot" until they entered hostile airspace or encountered flak.

As is common with American servicemen, some enterprising members of a combat crew assigned to 15th Air Force designed a field modification for the M1 helmet. Essentially, they used a screw-jack to force the sides of a standard M1 helmet shell farther apart. They then cut sections out of the side of fiber Hawley liners to accommodate the communications earphones. Other units modified M1 helmets by hammering pockets into the sides for headphones. These practices spread throughout the Army Air Force until aircrew-specific anti-flak helmets became available.

This helmet, found in Europe after the war, is a good example of the lengths aircrew members went to in search of protective armor. The sides of this M1 have been hammered out to allow earphones to be worn beneath the helmet. (Photo courtesy of David B. Powers)

The M3 Anti-Flak Helmet

During the early months of 1943, a prototype anti-flak helmet was unveiled and designated the T2. The T2 was a hybrid of the M1 but with some obvious differences. Like the field-modified M1 liners, the steel shell of the T2 was cut away to make room for earphones to be worn. In an effort to provide protection to these exposed areas, the T2 featured hinged ear flaps positioned over the earphones. They extended below the helmet's sides. Unlike the M1, the T2 did not use a separate liner, but rather had the suspension system mounted directly to the interior of the steel shell.

Welded to the exterior bottom edge of the earflaps was a hinged helmet loop, identical to the loops on M1 helmets. Like the M1, a khaki canvas webbing chinstrap was permanently sewn on. It served the obvious purpose of keeping the helmet in place plus held the earflaps down and in position. Instead of a hook system to hold the chinstrap together, the T2 chinstrap used a male/female snap system.

An unusual feature of this helmet was that after the helmet was painted olive-drab, the exterior was coated with flock, a cotton-lint material which looked like a coat of velvet. It prevented the steel shell from freezing to flyer's bare skin at high altitude and reduced static electricity.

In December 1943, the T2 was approved for aircrew use and redesignated the M3. Made from Hadfield Manganese, just over 210,000 of the three-

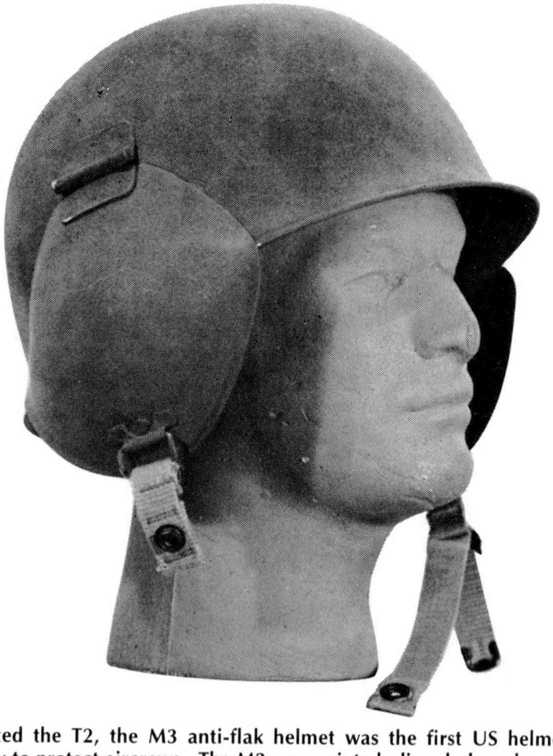

Originally designated the T2, the M3 anti-flak helmet was the first US helmet of WWII designed specifically to protect aircrews. The M3 was painted olive drab and was covered in flock, a cotton lint material to prevent the helmet from freezing to the skin at high altitudes.

pound, three-ounce M3 helmets were manufactured by McCord Radiator and Manufacturing Company of Detroit.

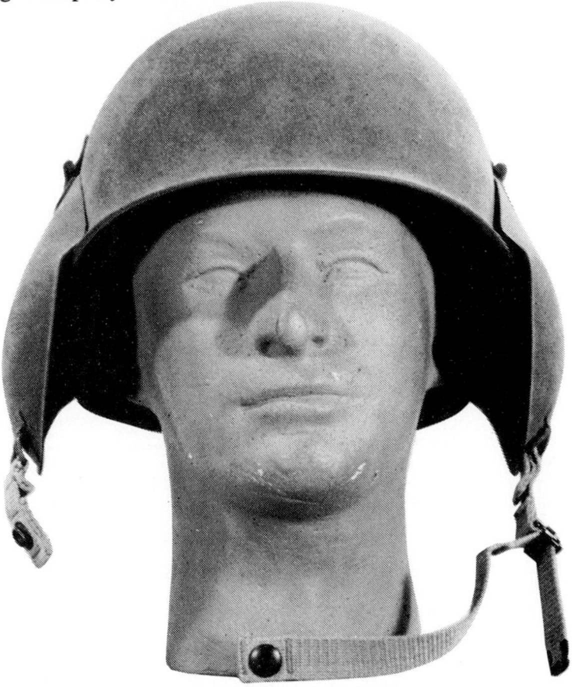

The M3 was worn on top of the flyer's leather or cotton flying helmet. With its hinged ear flaps, the M3 permitted flyers to wear earphones with improved comfort and protection. The ear flaps are an obvious throwback to the WWI experimental models 14, 14A and 15.

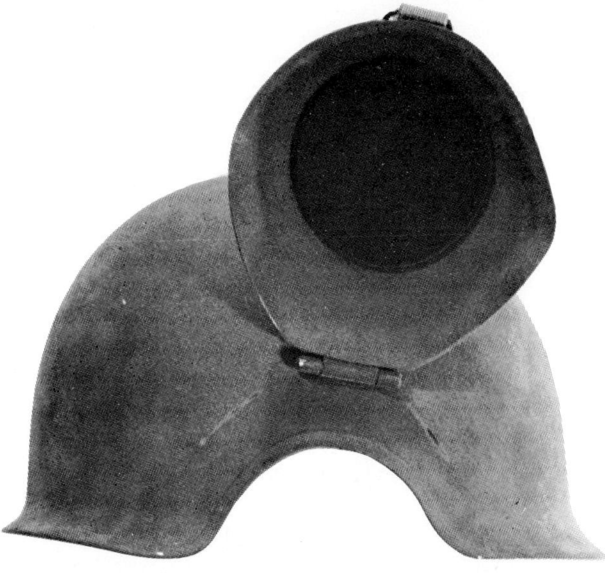

Essentially a hybrid of the M-1, the sides were cut away to allow earphones to be worn under the helmet. Like the M1, the M3 was made from Hadfield Manganese steel. A round brown felt pad was glued inside the ear flaps to reduce the noise and impact of the flaps against the earphones.

Unlike the M1, the suspension of the M3 was built into the shell. A steel band ran around the interior of the helmet and the suspension was clipped into it. The suspension could be easily removed from the helmet.

Slotted stud pins were riveted to the cotton herringbone twill suspension. To install the suspension, the pins were pushed down into the notch-anchor points on the steel band. The tension of the suspension coupled with the grooves on the pin was enough to hold the suspension in place.

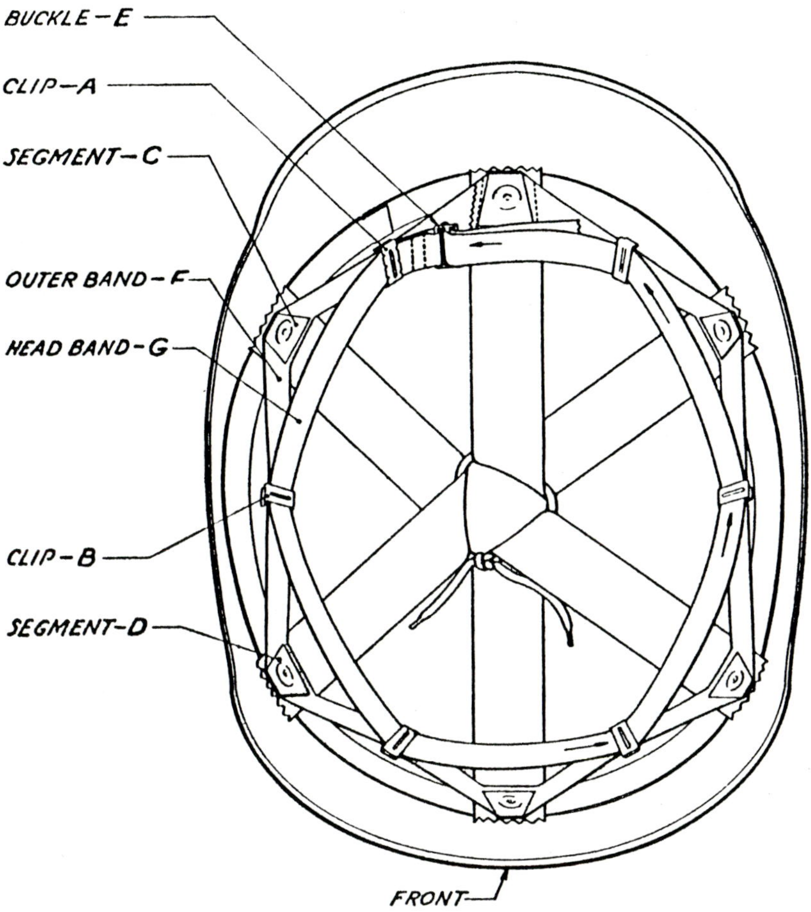

A copy of this diagram was supplied with each M3 describing how to adjust and remove the suspension from the liner.

An M3 suspension with its six slotted studs. Unlike the M1 helmet, the M3 suspension did not have a separate headband. The helmet was very large so that aircrew members could wear the leather or cotton flight helmet beneath it.

The M3 helmet used the same hinged chinstrap loops as the M1 helmet. A khaki webbed chinstrap was used to hold the helmet on the head. This photo shows the short end of the chinstrap, located on the right flap of the helmet. A snap was used to hold the chinstrap ends together.

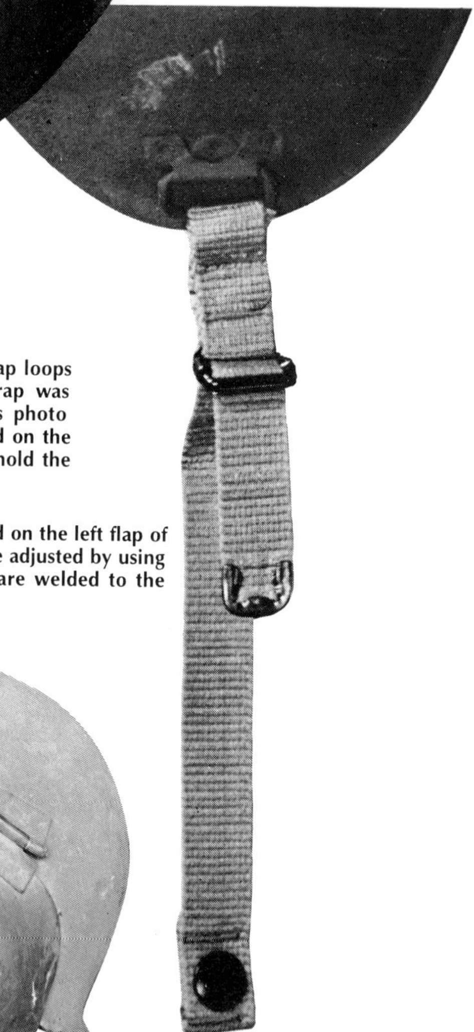

This is the longer section of the chinstrap located on the left flap of the helmet. The length of the chinstrap could be adjusted by using the sliding buckle. Notice the chinstrap loops are welded to the outside of the ear flaps on the M3.

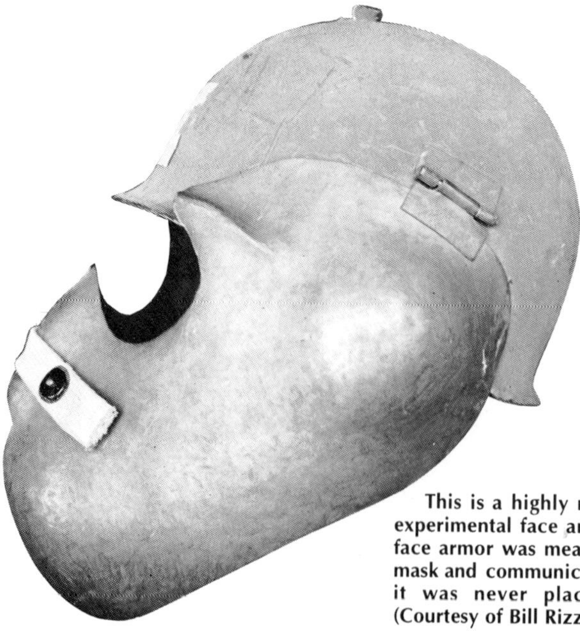

This is a highly modified M3 helmet supporting experimental face armor designated T6E1. This full face armor was meant to be worn over the oxygen mask and communications earphones. While tested, it was never placed into operational service. (Courtesy of Bill Rizzutti)

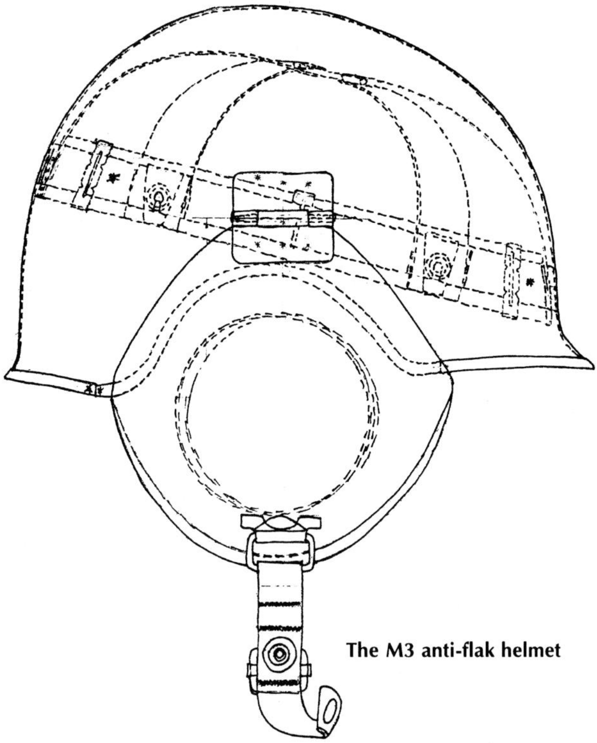

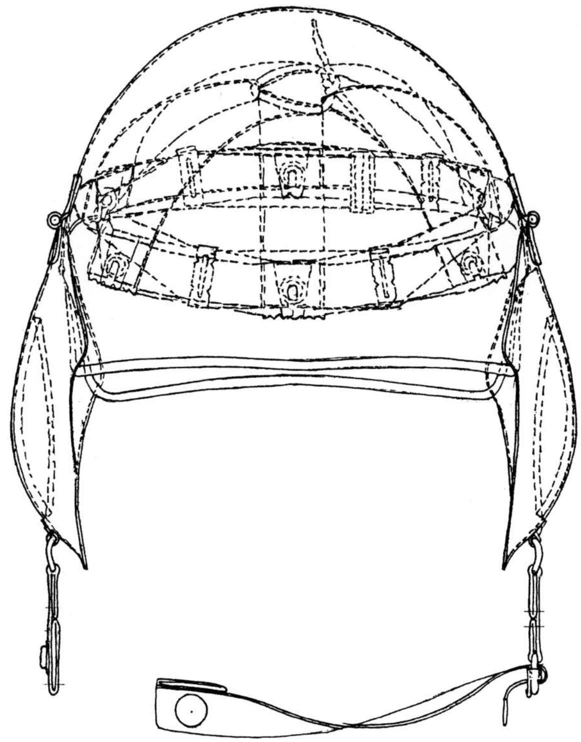

The M3 anti-flak helmet

The M4 Series Anti-Flak Helmets

For pilots, navigators, and door-gunners, modified M1 helmets or M3 helmets greatly improved their chances of surviving an encounter with flak. However, many positions in aircraft, such as cramped upper turrets and lower ball-turrets, were so small that it was impossible to wear a helmet the size of an M1. The main problem was the helmet sat too high on the head.

The Chief Surgeon of the Eighth Air Force, Colonel Malcolm Grow, was determined to find a solution to the dilemma and developed an anti-flak helmet specifically for aircrew members who occupied the tight confines of turrets.

His initial design weighed just over two pounds and looked like a skull cap. Unlike the M1, the "Grow helmet," as it was often called, did not have an internal suspension. It rested directly on the leather or cloth flying helmet.

A prototype was submitted to the Ordnance Department for consideration and was designated as the T3. Tests concluded that while the helmet did not have the protective qualities of the M3, it did improve survivability.

The helmet was redesigned as the M4 and placed into service.

The M4 anti-flak helmet. While many aircrew members benefited from the M3 helmet, several crew positions such as ball turret gunners didn't have room to wear such a large helmet. The M4 was designed to fill that gap.

The M4 was invented by Army Air Force Colonel Malcolm Grow. Weighing in at just over two pounds the M4 was designed to be worn on top of the leather flying helmet.

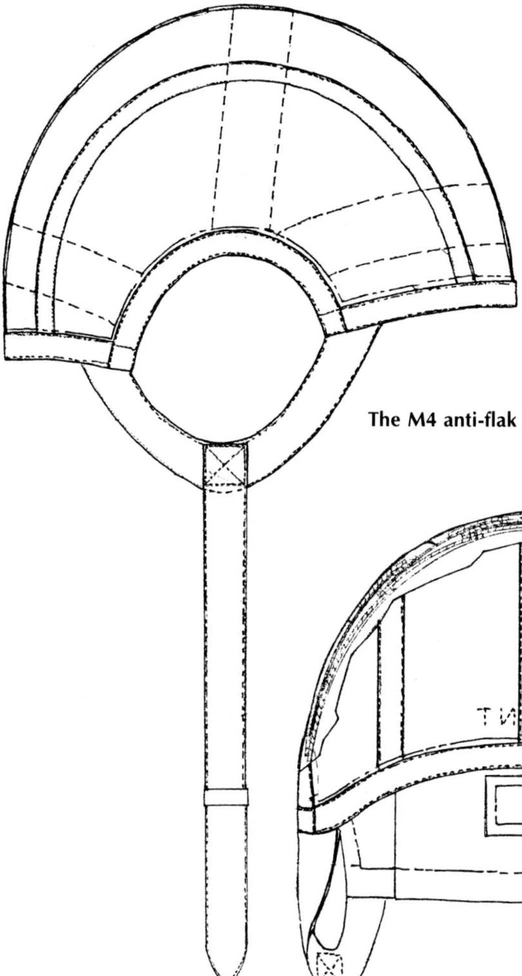

The M4 anti-flak helmet

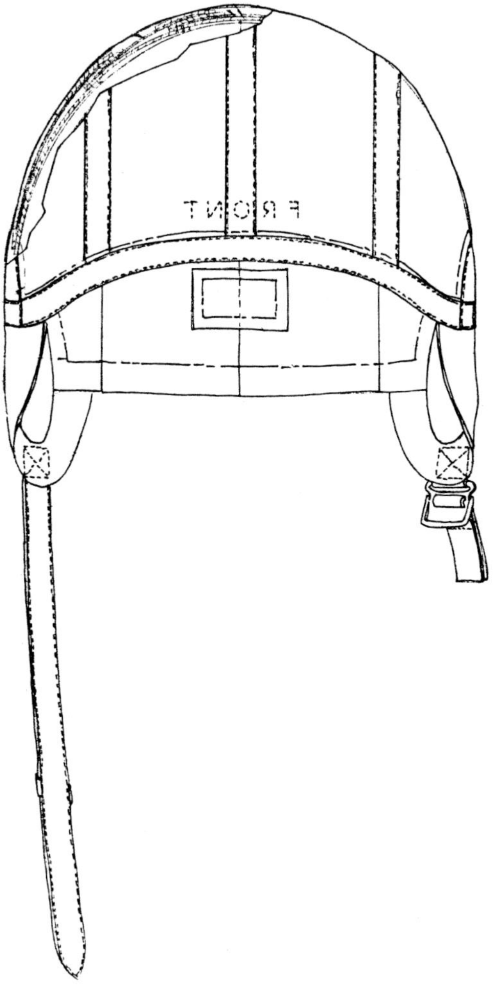

The M4 was manufactured in two versions: One in England and the other in the United States.

The British-made M4's were constructed of overlapping pieces of Hadfield Manganese covered in a very dark brown leather. The interior of the helmet was covered in a chamois lining. Since the helmet had cut-outs for the ears, the chinstrap circled the ear holes and used a lift-the-dot fastener on the right side and an adjustable buckle fastener on the left. These were made by Wilkinson Sword Company.

Also using Hadfield Manganese, American-made M4's were very similar to the British model except, instead of leather, the exterior covering material was an olive-drab twill. The chinstrap on the American model was leather but it used a buckle rather than a lift-the-dot fastener.

Initially designated the T3, the M4 was made in both Britain and the US. British examples were made by Wilkinson Sword Company and were covered in a dark brown leather. U.S. M4s were covered in an OD cotton twill material.

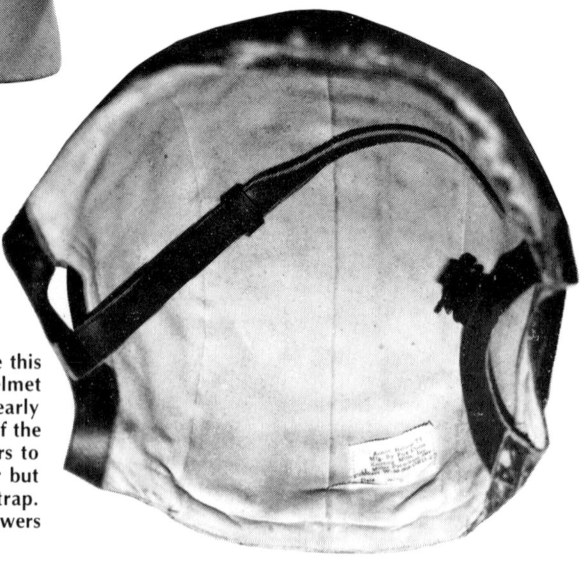

The nomenclature tag inside this example reads Armor, Helmet T3. This appears to be an early attempt at U.S. production of the helmet as the helmet appears to be covered in a dark leather but has the U.S. style of chinstrap. (Courtesy of the David B. Powers Collection)

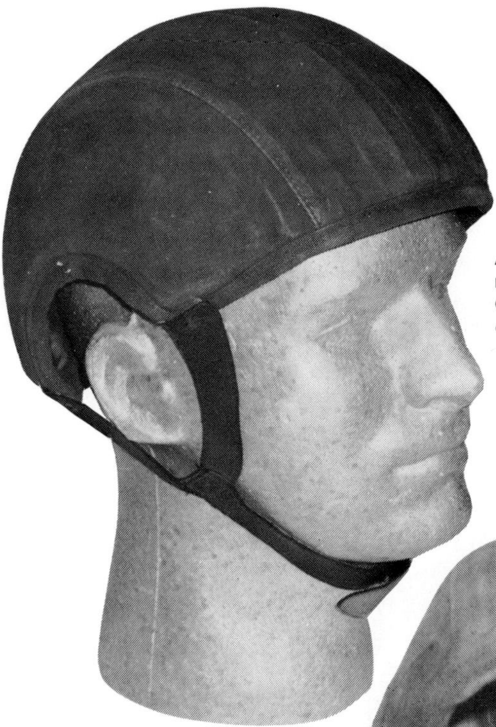

An exterior view of an American-made M4 anti-flak helmet with its OD twill cloth covering. (Courtesy of Col. Barry Smith)

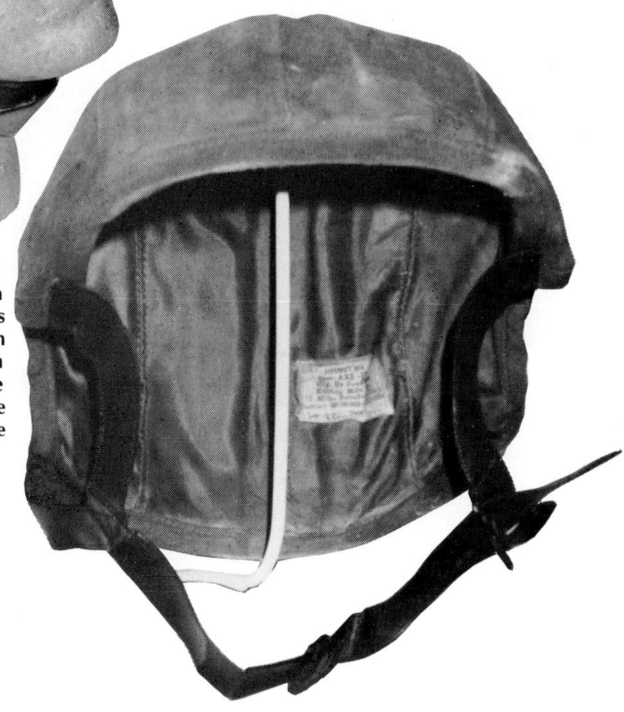

From the front a view of an American-made M4 helmet. This chinstrap is much different from the British version of the helmet in that rather than snap in place, the American version used a buckle mechanism. (Photo courtesy of the David B. Powers Collection)

In 1944, a new model of the M4 was created after a determination that armored ear flaps would improve the helmet's protective qualities. Designated the M4A1, these helmets were made with ear flaps sewn into heavy olive drab cloth covers. They were attached to the helmet with a heavy cloth hinge. Another feature of the M4A1 was the addition of tabs positioned across the top of the helmet to secure goggles in place. Many M4 helmets were converted to M4A1's by using a pre-fabricated cover which could be slipped over the M4 body.

The chinstrap on the M4A1 was repositioned to the bottom of the ear flaps, similar to the M3.

This profile photograph shows a good view of the helmet's cut-outs for the ears and the chinstrap assembly. While the left side of the chinstrap buckles to the helmet, the right side uses a press-the-dot fastener to secure the helmet in place. The leather chinstrap shown was used only on the British version of the M4.

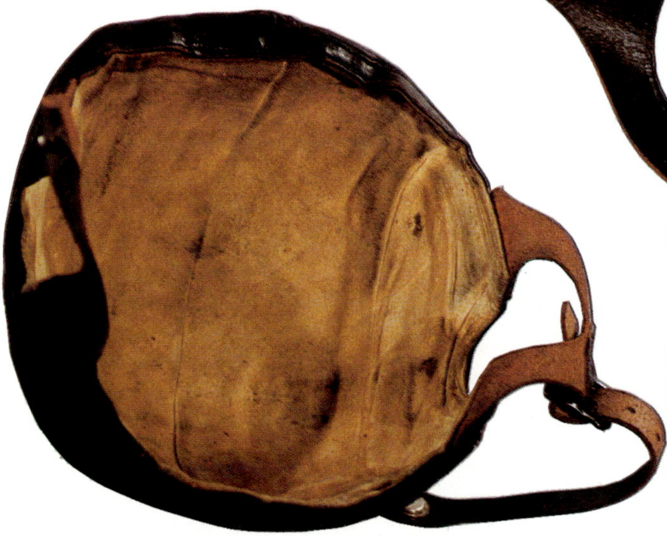

A look inside a British-made M4 helmet. The interior was covered with a chamois lining. Notice the helmet has cut outs for the ears. M4 helmets relied on overlapping strips of Hadfield Manganese for protection.

In 1944, the final helmet of the M4 series was born. In an effort to improve its fit, it was a slightly longer helmet. Designated the M4A2, it would replace all earlier M4 variants. The M4A2 was constructed from Hadfield Manganese and covered with a heavy olive-drab cloth covering. Weighing in at two pounds, 12 ounces, the M4A2 was lined with a smooth, dark, olive-drab material.

The steel ear flaps on the M4A2 were covered inside with the same olive drab cloth as the main helmet body. The flaps were secured to the helmet by using a heavy canvas hinge.

The chinstrap on the M4A2 was made of a light brown leather. The right end was sewn to the exterior of the ear flap. On the left ear flap a short leather tab with a buckle was sewn on. The chinstrap held the helmet and ear flaps in place.

The final variant of the M4 series of anti-flak helmets was the M4A2. This helmet was slightly longer than both the M4 and the M4A1. It weighed two pounds, 12 ounces and like most U.S. helmets, used Hadfield Manganese.

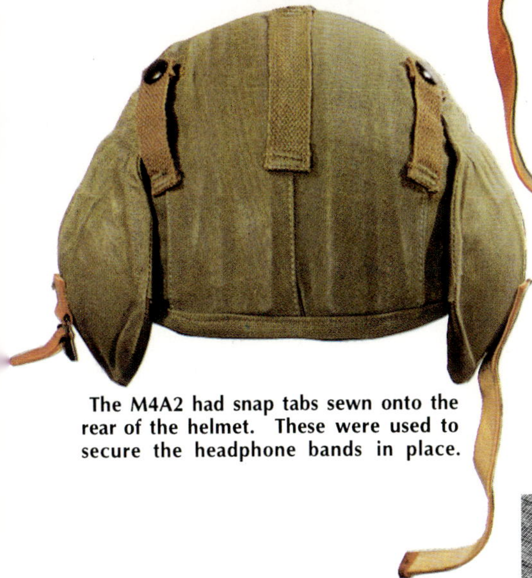

The M4A2 had snap tabs sewn onto the rear of the helmet. These were used to secure the headphone bands in place.

The M4A2s were made exclusively in the U.S. and were covered in the same OD twill material as used on U.S.-made M4 helmets. The chinstrap on M4A2s was constructed of leather but was sewn in place on the right side. The chinstrap was secured with a buckle sewn onto the left-side flap.

A typical example of a maker's tag inside an M4A2 helmet.

Across the top of the helmet, snap-tabs were positioned to hold straps from goggles or masks in place.

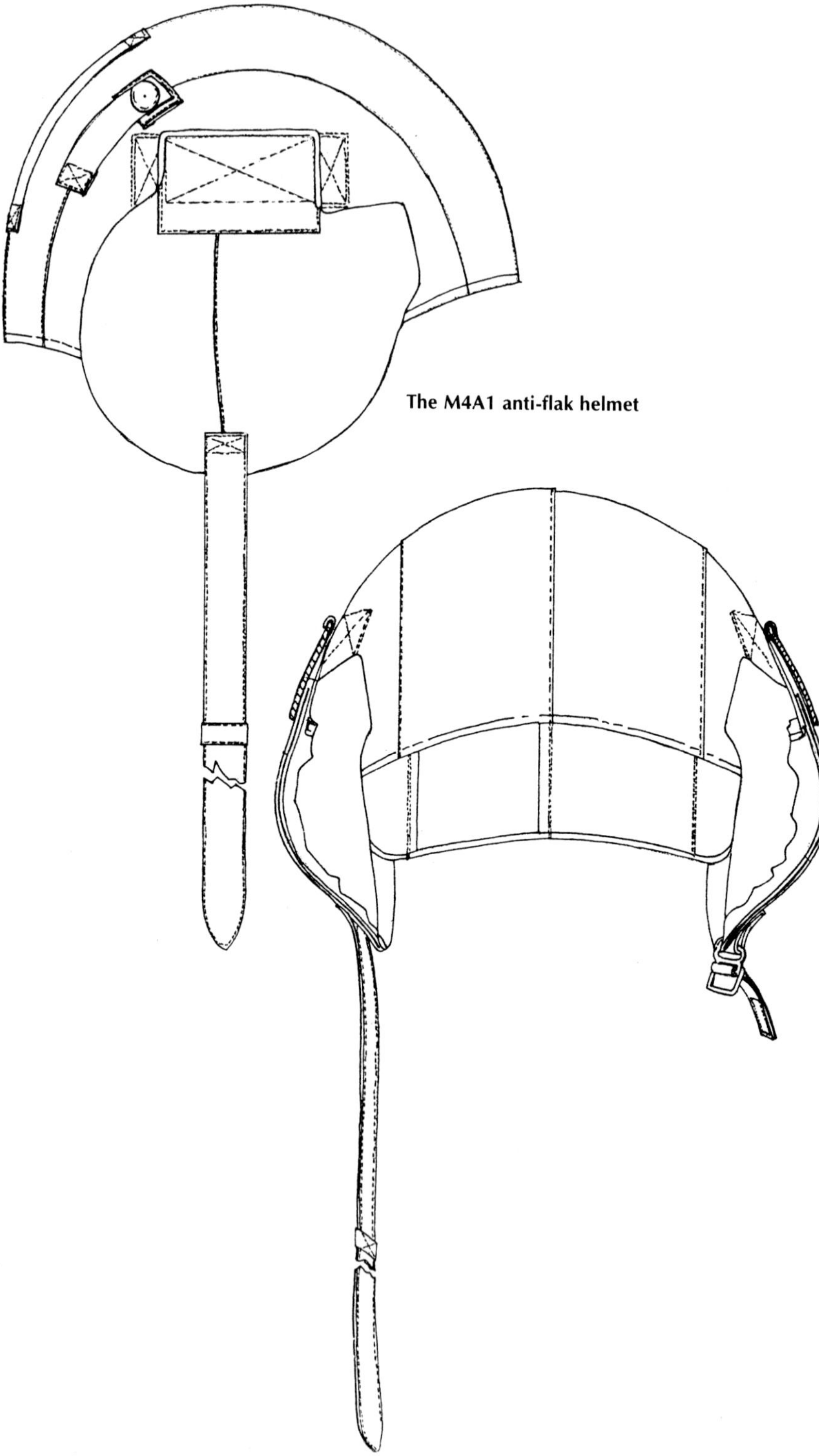

The M4A1 anti-flak helmet

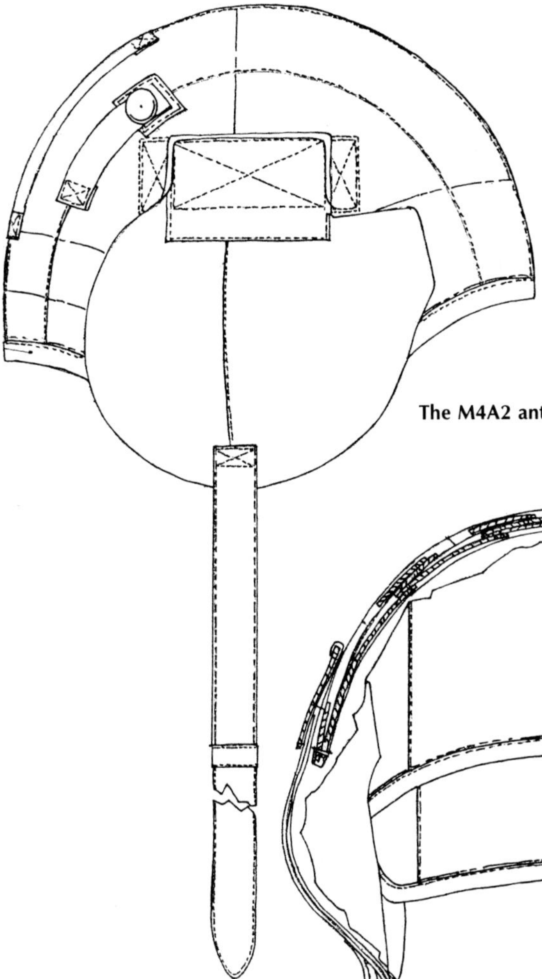

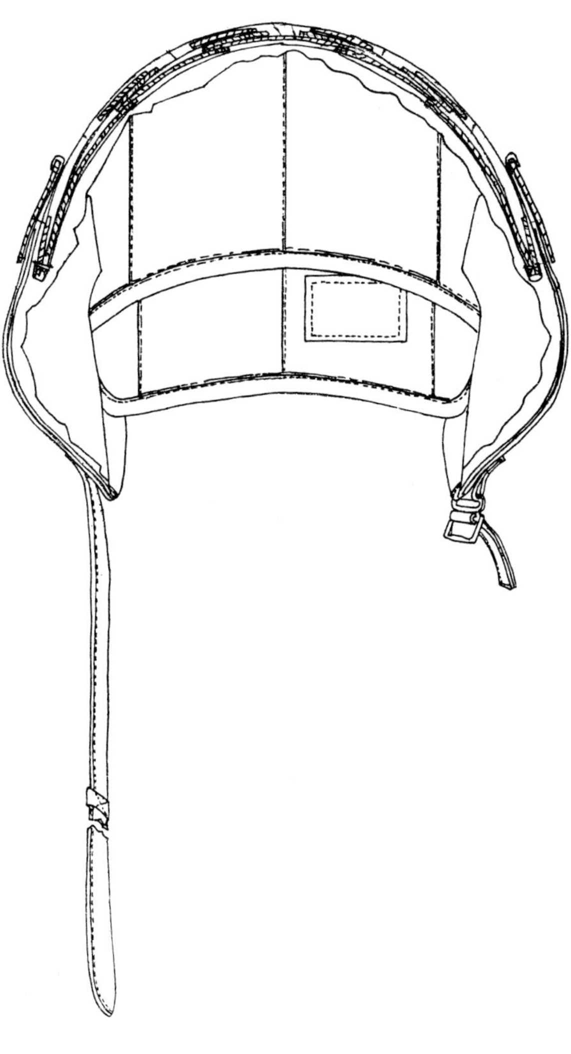

The M4A2 anti-flak helmet

The M5 Anti-Flak Helmet

Having different models of similar equipment always annoyed the Ordnance Department. Since 1942, efforts to develop a standardized anti-flak helmet for all aircrew members had been progressing. The most promising design took elements from both the M1 and the M3.

This standardized helmet, known as the M5, was seen as the solution to the variety of helmets in use around the Army Air Force. The helmet, made from Hadfield Manganese, continued the practice of attaching ear flaps to the steel shell.

Each flap was riveted to one side of a hinge. The other side of the hinge was riveted to the helmet body. What made the M5 such an improvement over earlier efforts was that it sat much lower on the head and could be worn in nearly every compartment on any bomber, cargo, or glider aircraft. The suspension of the M5 was secured to a band of steel welded to the inside of the helmet shell. Unlike the M3, the M5 did not have a flair to the front.

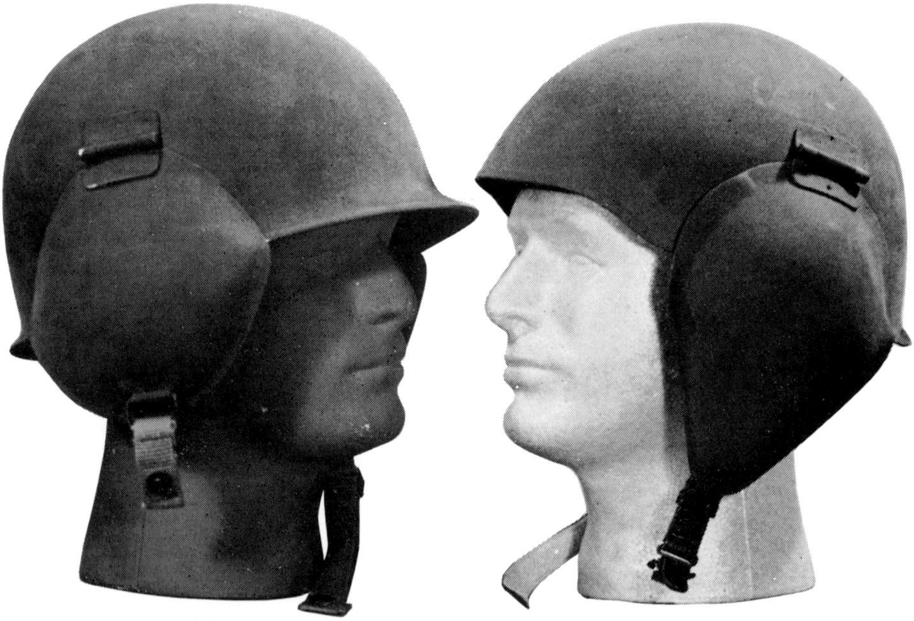

A comparison of the M3 helmet (left) and the M5 helmet (right). The M5 was the final anti-flak helmet adopted for service during the Second World War.

A major innovation which would influence future helmet production was the method used to attach the chinstrap to the helmet. Like the M1 and the M3, steel hinged chinstrap loops were welded to the helmet. On the M5, these loops were welded on the inside bottom of each ear flap. Instead of sewing the leather chinstrap to these loops, a clamp called the T1 was developed. This clip fit through the chinstrap loop and clamped around the end of the chinstrap. The chinstrap could still swivel back and forth and was very secure. The key element was the ability to easily remove the clamp and the chinstrap if it broke or became damaged.

The chinstrap, when coupled together, was held in place by a fold-over buckle clamped against a small row of teeth.

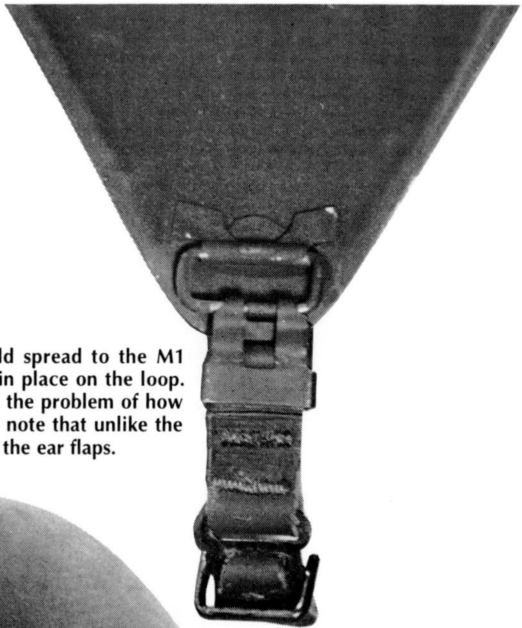

An innovation from the M5 which would spread to the M1 helmet was the clip holding the chinstrap in place on the loop. This was the T1 clamp. This clamp solved the problem of how to replace lost or broken chinstraps. Also note that unlike the M3, the chinstrap loops are welded inside the ear flaps.

The M5 helmet took the best elements of both the M1 and the M3 helmets. The M5 was designed to replace all existing anti-flak helmets. Like the M3, the M5 used Hadfield Manganese and hinged flaps to provide increased protection, while allowing earphones to be worn.

The M5 was coated in the same flock material as the M3. Nearly 100,000 of these nearly three-pound helmets were manufactured in 1945.

For the most part, the end of World War II signaled the end of the anti-flak helmets. While used in small numbers during the Korean conflict, the advent of jets reduced the threat of flak and small arms fire to a degree that they were no longer considered necessary.

The M5 sat lower on the head than the M3 and thus took up far less space. This permitted it to be worn in cramped turrets where the M3 could not be used. The M5 was also covered in flock to prevent freezing at high altitude.

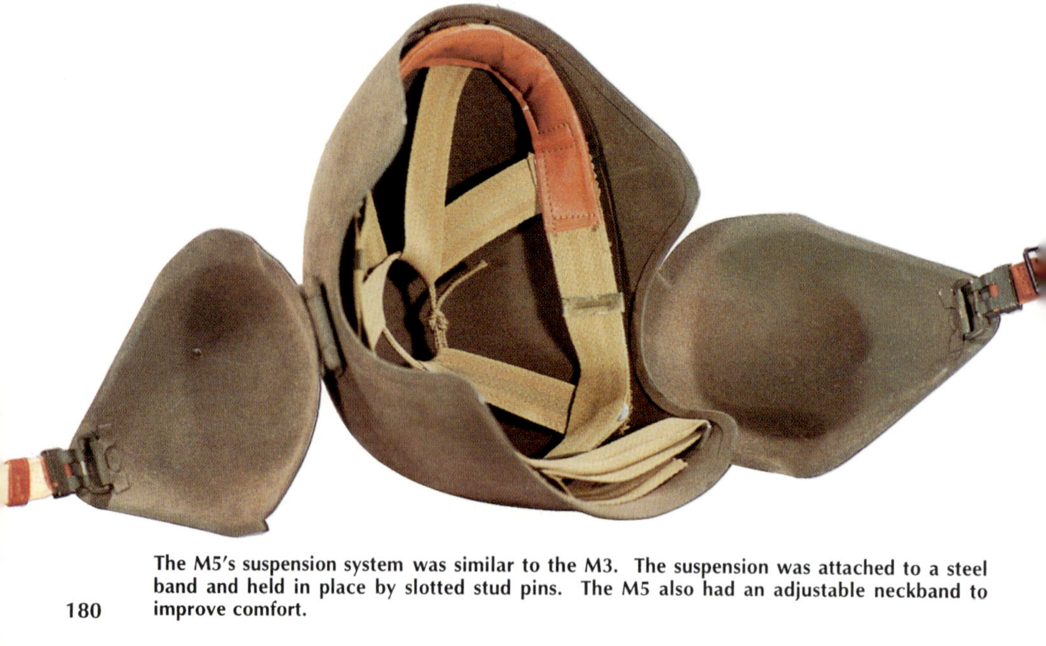

The M5's suspension system was similar to the M3. The suspension was attached to a steel band and held in place by slotted stud pins. The M5 also had an adjustable neckband to improve comfort.

A closer look at the adjustable headband in the rear of the M5 helmet.

A look at the suspension system of the M5 helmet. Unlike the M3, the front of the M5 suspension is covered in leather. The M5 suspension uses six anchor points to affix it to the steel band inside the helmet shell.

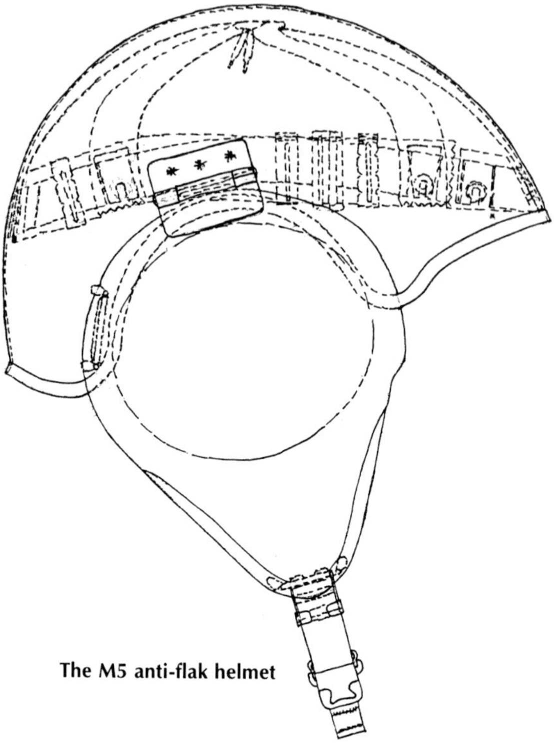

The M5 anti-flak helmet

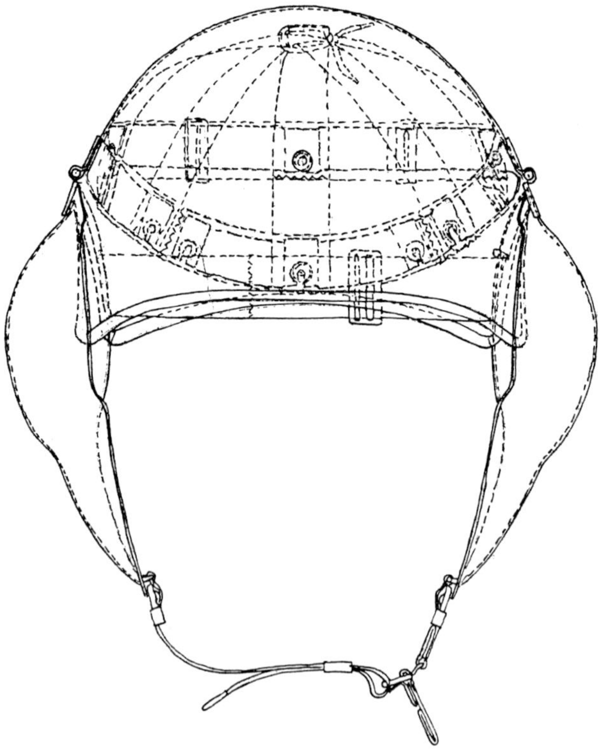

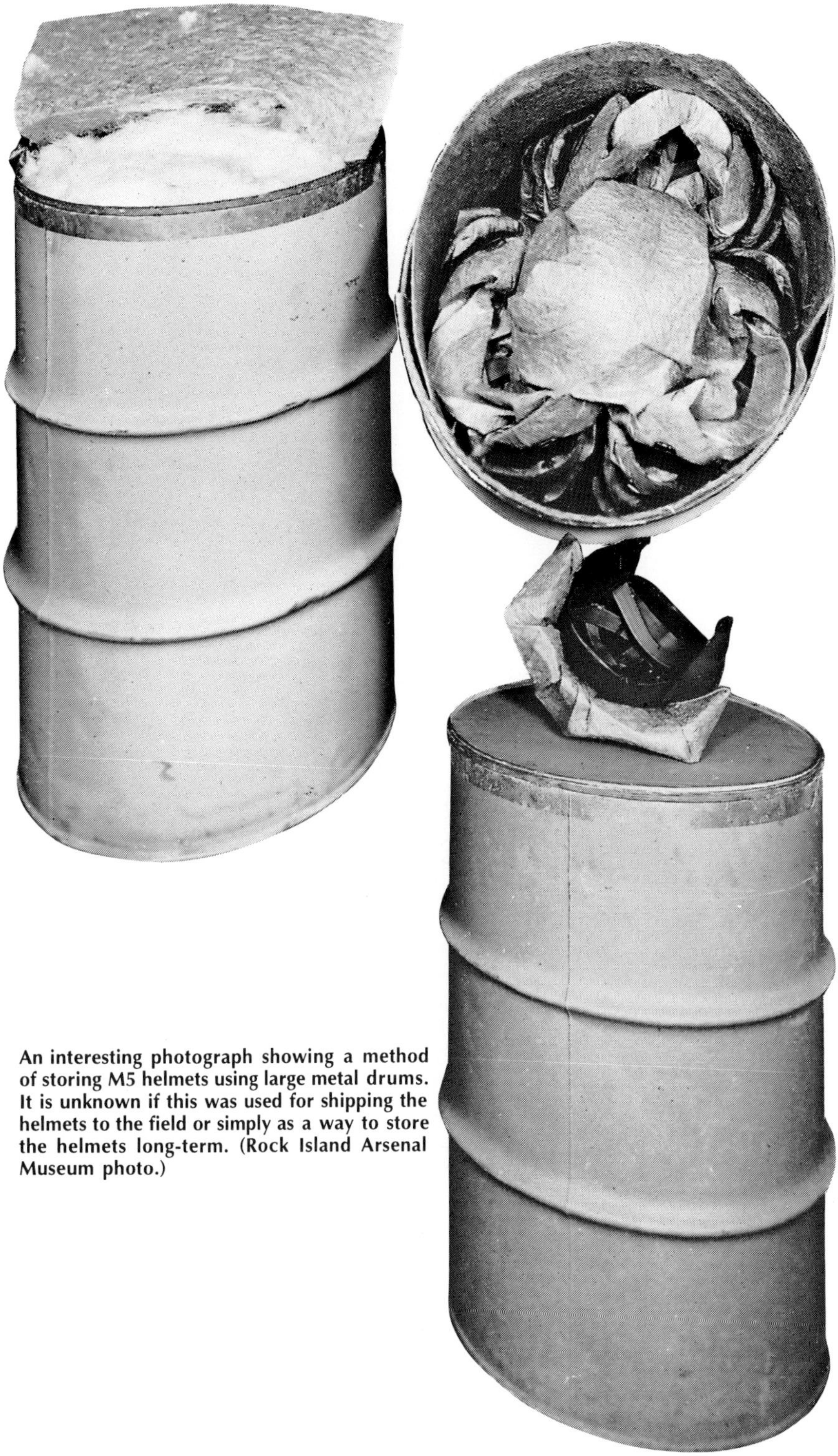

An interesting photograph showing a method of storing M5 helmets using large metal drums. It is unknown if this was used for shipping the helmets to the field or simply as a way to store the helmets long-term. (Rock Island Arsenal Museum photo.)

The most mysterious of all World War II American anti-flak helmets is so obscure it doesn't even have a designation. The product of Project H-28-CL, the name of the helmet and its strange yellow star have been lost to history. The helmet is very similar to the M4A2 helmet and may have been an early prototype designed to replace the M4 helmet. Upon the decision to use the OD twill canvas could the H-28-CL project have been dropped? What of the yellow star? Could it have been applied simply as an expeditious way for armor designers to identify it and evaluate examples which made it to the field for testing? (Courtesy of the Tom Gordon Collection)

The nomenclature tag of the helmet designed during project H-28-CL. (Courtesy of the Tom Gordon Collection)

CHAPTER SIX

KOREA, VIETNAM, AND BEYOND. THE POST WWII M1 HELMET

The end of World War II did not signal the end of reliance upon the M1. On the contrary, the helmet would continue to be refined and improved and would remain in use well into the 1990's.

Almost immediately after the war, the military began to study the effectiveness of the M1 and its ability to protect the people who wore it. A secret post-war study of casualty data, Project ALCLAD, indicated that 70 percent of the total wounds to ground forces were caused by fragments and the remaining 30 percent were by bullets and other causes.

These reports indicated that the head was twice as likely to be struck by a projectile in comparison to other parts of the body. The reasoning for this trend was a belief the head was a preferred target and was often more exposed to enemy fire.

The Operations Research Office, operating from Johns Hopkins University, collected and evaluated wartime data. Based on the data, they determined the M1 prevented serious or fatal wounds in more than 76,000 men.

Project ALCLAD recommended "that the provision of armor helmets to all ground combatants in the event of war be continued as a military policy." Other recommendations by the project stressed the need to constantly develop new helmets with a goal of affording increased protection, especially to the cranium and forehead region.

Taking these recommendations into consideration, the Army quietly began work on developing a new helmet for American forces.[1]

In the meantime, the battle-tested M1 would continue to serve the military.

After World War II, the Army set about the task of reclaiming the equipment and materiel issued to armies and individuals. This task obviously included reclamation and assessment of helmets. An undated tentative specification from the Jeffersonville Indiana Quartermaster gives some interesting insight into the reclamation process.[2]

---

[1] *See Chapter 7.*

[2] *Authors note: From detailed reading of the specification, I believe it is from 1945 or 1946. My reasoning is this specification was designed for contractors to bid for accomplishing the reclamation process. It also references some late-war items such as the M5 helmet and hook-and-ball connectors.*

INSPECT FOR	DEFECTS ACCEPTABLE IN COMBAT SERVICE-ABLE WITHOUT REPAIR	DEFECTS ACCEPTABLE IN REPAIR FOR COMBAT SERVICEABLE
Dents	Scarcely perceptible dents are acceptable provided this is the only defect. Dents in excess of 1" diameter and having a depth of more than 1/4" will not be acceptable where more are concentrated in an area less than 3" in diameter, or in such a manner as to affect the contour of the helmet.	All dents shall be removed and the helmet refinished.
Cracks	Cracks of any kind are not acceptable.	Cracks shall not be repaired.
Scoring and Scratches	Deep scoring or scratches in the metal which would cause cracks or holes to form will not be acceptable. Scratches which cause the metal to reflect light will be acceptable for repair.	Scratches which reflect light shall be repaired and the helmet refinished.
Burns	If there is any evidence that heat has been applied to the helmet causing discoloration or weakening of the metal the helmet will not be acceptable.	Burns shall not be repaired.
Holes	Holes will not be acceptable with the exception that small holes caused by D-Ring removal will be acceptable for repair.	Holes caused by D-Ring removal shall be repaired.
Welds	If there has been an indication that the helmet has been welded during service, it will not be acceptable.	Welds shall not be repaired.
Loose Rims	If the spotwelding holding the rim has been destroyed, or the rim has become detached, the helmet will not be acceptable.	If the rim is loose but has not become detached, and the spotweld holding the rim is intact, it shall be repaired and the helmet refinished.
Cleanliness	Non-deteriorating spots, stains, and dirt will be acceptable after helmet is brushed clean, provided this is the only defect.	Spots, stains and dirt which have caused deterioration of the finish shall be repaired and the helmet refinished.
Hardware	Loose, bent, split or missing clips will not be acceptable. Rusty clips or clips having the protective finish destroyed will not be acceptable.	Loose, bent, split, or missing clips will not be acceptable. Rusty clips or clips without a protective finish shall be replaced.

INSPECT FOR	DEFECTS ACCEPTABLE IN COMBAT SERVICE-ABLE WITHOUT REPAIR	DEFECTS ACCEPTABLE IN REPAIR FOR COMBAT SERVICEABLE
Buckles, tongueless bar, with hook	Bent, split, broken or missing buckles will not be acceptable. Rusty buckles or buckles having the protective finish destroyed will not be acceptable.	Bent, split, broken, missing or rusty buckles, or buckles without a protective finish shall be replaced. Buckles having small spear-head bent, broken or destroyed shall be replaced.
Hooks	Old style wire hooks are not acceptable. New style spring hooks that are bent, split, broken or missing are not acceptable. Hooks having the protective finish destroyed are not acceptable.	Old style hooks shall be replaced by new style spring hooks. Bent, split, broken, rusty or missing hooks, or hooks without a protective finish shall be replaced.
Tongue-Ball	Bent, split, broken, rusty, flattened, or missing ball tongues are not acceptable. Ball tongues having the protective finish destroyed will not be acceptable.	Bent, split, broken, rusty, flattened, or missing ball tongues without a protective finish shall be replaced.
D-ring	Old Style looped wire D-Rings are not acceptable. New style hinged loop D-Rings that are bent, split, or broken, are not acceptable. D-Rings having protective finish destroyed are not acceptable.	Old style looped wire D-Rings shall be replaced by new style hinged loop D-Rings. Bent, split, broken or rusty D-Rings shall be replaced. D-Rings without a protective finish shall be refinished.
Webbing	Breaks, cuts, torn places, or excessively worn areas are not acceptable.	Webbing shall not be repaired.
Bar tacks The stitching used to sew chin strips onto the loops.	All bar tacks shall be intact and free from defect.	Bar tacks having thread missing or containing rotten thread shall be repaired provided tacking is the only defect.

As helmets were turned into the Quartermaster Depots, they were inspected to determine if the helmet required any repair, or if they were to be disposed of as unserviceable. The inspection guide inspected for, dents, cracks, burns, scratches, holes, broken welds, loose rims, chinstrap loops, chinstrap and liner hardware, hooks, and webbing.

It is interesting to note that the oldest M1 helmets with the welded chinstrap loops were no longer considered acceptable. The removal of the chinstrap loops, or D-Rings as the checklist calls them, was accomplished by shearing the loop off as close to the helmet as possible. The new style hinged helmet loop was then spot welded in its place.

The chart also indicates helmets were refinished. The process included soaking the helmet in a caustic bath to remove dirt and the old finish. The helmet then had any welding improvements made, before being primed and repainted.

The fact that helmets were restored or refinished at depots and installations around the world would account for the wide variation in M1 helmet colors and textures after World War II.

The specification also points out that defective chinstraps were removed and replaced with new chinstraps which were bar-tacked or sewn onto the helmet chinstrap loops. This is an indication that the use of the M1 helmet chinstrap clip was not in widespread use at the end of the war.

1950s Modifications and Alterations

From the end of WWII until the start of the Korean War, very little was done in the way of upgrades for the M1 helmet.

With the exception of a 1951-dated specification diagram, there is not much evidence to show that M1 steel shells were produced during the Korean War. The 1951-dated specification indicates several potential production changes were being considered in the steel shell.[3] One such change shown in the 1951 specification diagram clearly is the helmet rim being joined at the rear of the helmet.

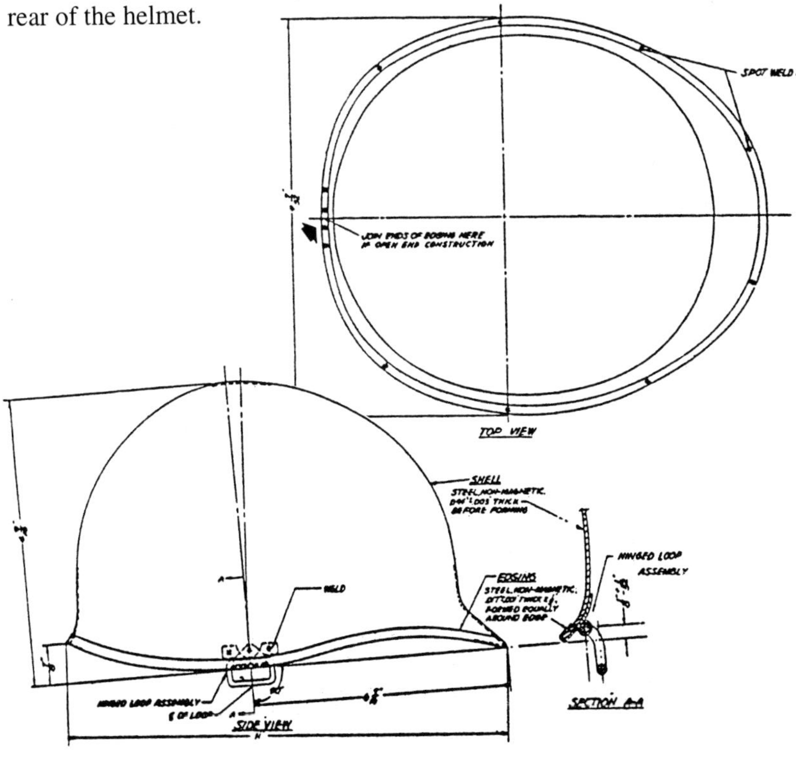

---

[3] *Authors Note: Try as I could, I was unable to locate any manufacturing or production data during the immediate post-WWII or Korean War era. Could this be due to the survival of an abundance of the 22,000,000 shells made during WWII? While a 1951 specification does exist, that in itself does not indicate a production run of shells.*

An Army photograph of the M1 helmet from the Korean War era. The primary difference between the M1 helmet of WWII and Korea was the chinstrap. The steel shell on this helmet sports the new T1 chinstrap clip and the T1 chinstrap attachment clamp. (US Army Photo, Courtesy of the Quartermaster Museum)

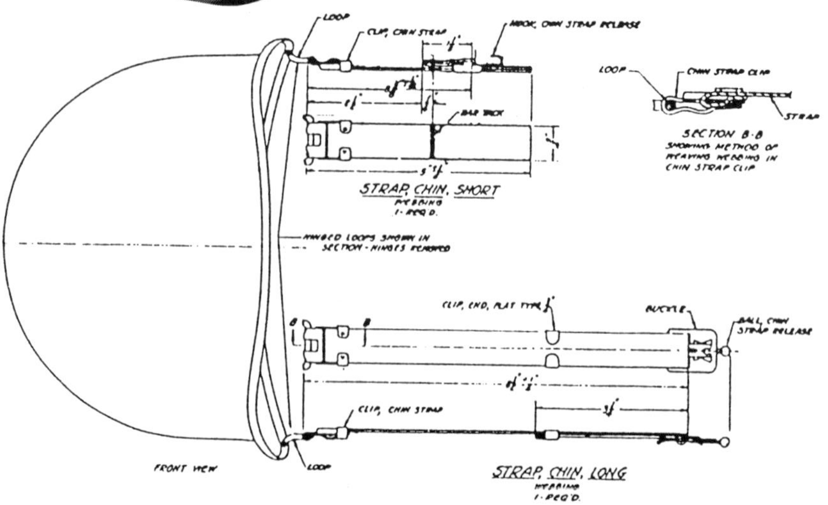

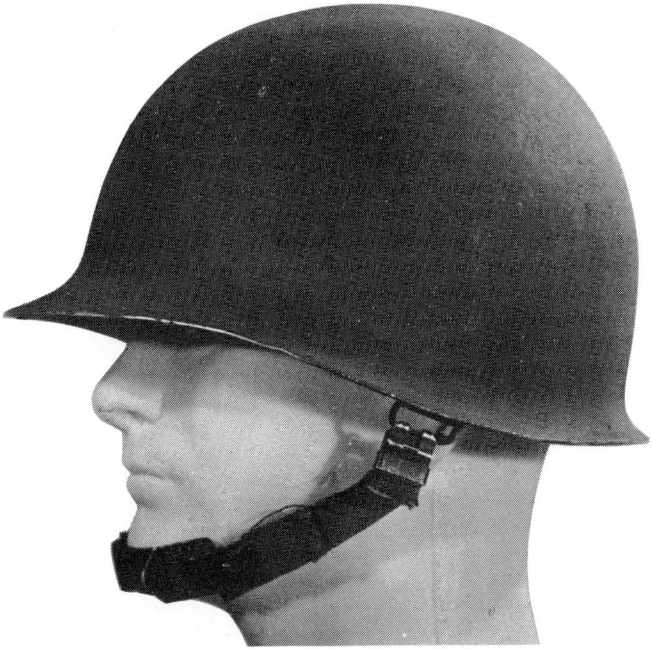

An M-1 helmet in typical Korean War era configuration. Slightly lighter olive drab, dark olive drab No. 7, chinstrap with T1 chinstrap hook and clip. Notice the cork texture applied in WW II is gone, replaced by a fine sand aggregate. Many WWII veteran M1's served in Korean with updated finishes and accessories.

While not much happened with the steel shell, in 1951, several accessory items developed during World War II became standard items. These included the T1 Chinstrap Release and the T1 Chinstrap Clamp. Other small changes included eliminating the olive drab painted adjustment mechanisms on the leather liner chinstrap in favor of flat black. This was also done with the headband installation clips.

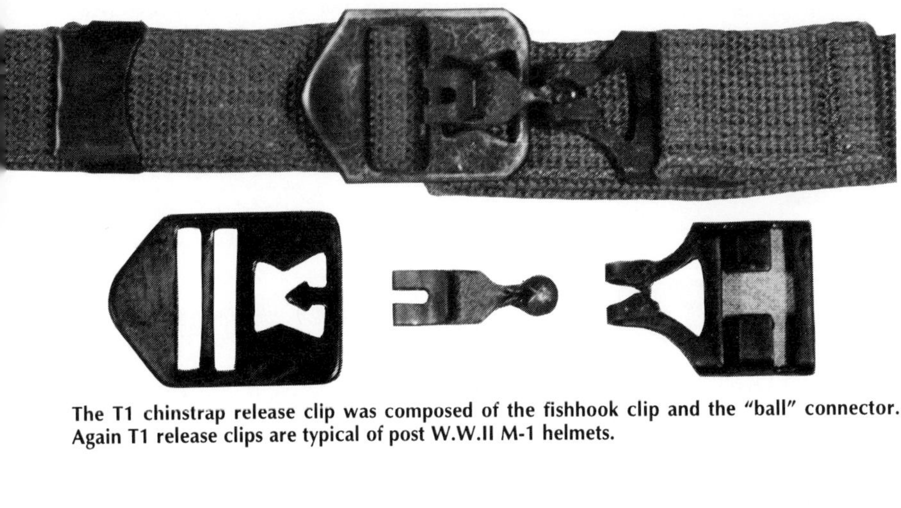

The T1 chinstrap release clip was composed of the fishhook clip and the "ball" connector. Again T1 release clips are typical of post W.W.II M-1 helmets.

The T1 chinstrap clamp was the typical method used to secure the chinstrap to the loops on Korean War era configured M-1 helmets. It was invented during WWII but saw very little use during that conflict. If the strap was broken or damaged the T1 clamp could be removed to allow the strap to be replaced or repaired.

The T1 clip was essentially a clamp which bent over the chinstrap loop and clamped onto the web chinstrap. The clamp could be removed by prying on it with a bayonet or screwdriver.

The dark OD No. 7 color web chinstrap most often encountered on post WWII M-1 helmets. Notice the T1 chinstrap connector clip.

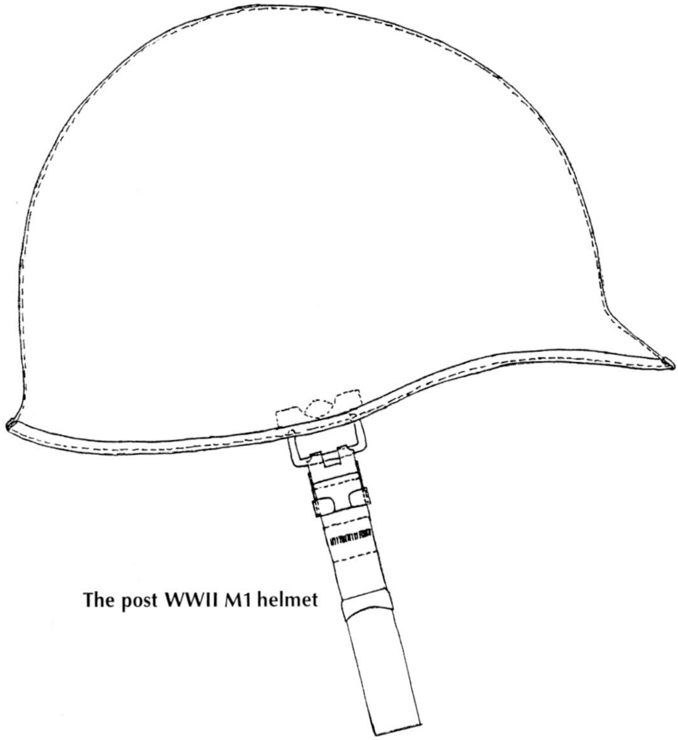

The post WWII M1 helmet

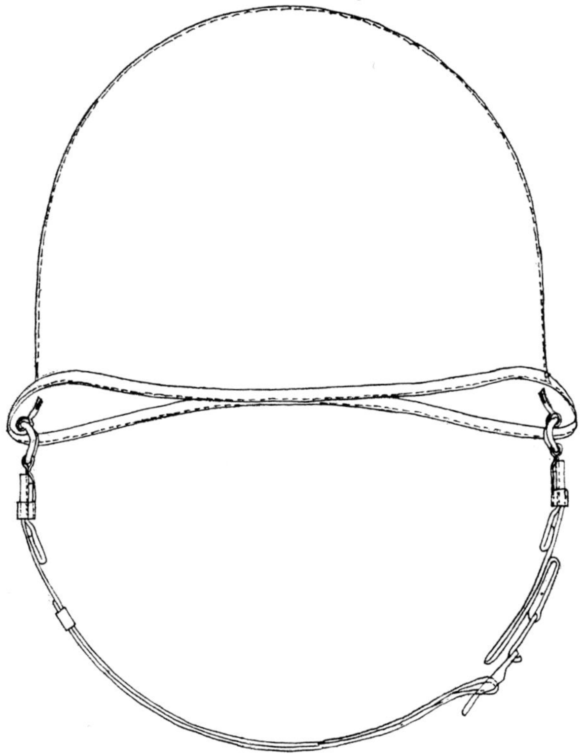

Corporal Charles Lovin at the time of the described action.

Rarely, a helmet can tell a story of individuals and events. The illustrated Korean War era M-1 helmet is such an example. On November 25, 1950, "C" Company, 1st Battalion, 23rd Regiment of the 2nd Infantry Division was sent up from reserve to occupy Hill 329, also known as "Chinaman's Hat." This strategic hill was situated about two miles north of Kujang-dong which lay deep in North Korea. "C" Company dug in next to the 61st Field Artillery.

That evening masses of Chinese soldiers hit the position and the Americans were thrown off the hill. On the morning of November 26 a counter-attack was launched to retake the important position. Participating in this attack was Corporal Charles Lovin of "C" Company. As he reached the crest of the hill, a young Chinese infantryman jumped up from the trench line about ten feet from him and fired off a shot. The bullet hit the illustrated helmet and passed through both the helmet and the liner. Lovin went down as the helmet slammed against his head. Miraculously he only suffered from a headache, ringing in both ears and slight wounds from bullet fragments which hit below his left eye and grazed his left ear. This action resulted in the awarding of his second Purple Heart.

Either Sergeant John A. Pittman, also with "C" Company, or Corporal J.P. Littrell from his squad shot the Chinese soldier moments after Lovin was hit. After a fierce engagement the hill was secured

Note: The 2nd Inf. Div. decals were sent to Korea from Japan and applied before this action.

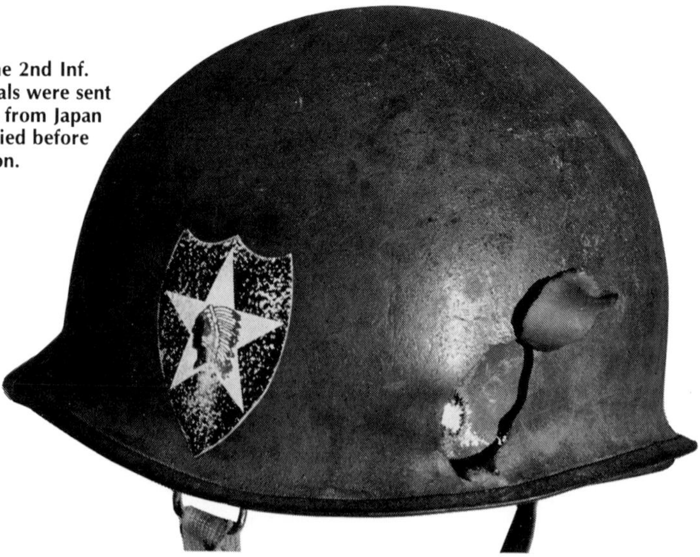

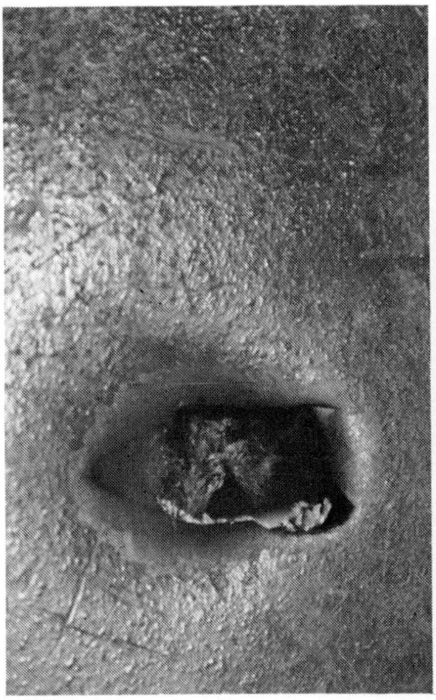

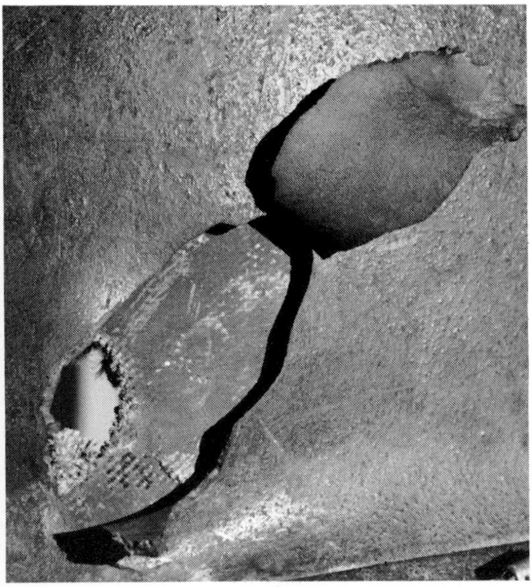

Entrance and exit holes.

with 410 of the enemy killed, 111 taken prisoner, and an unknown number of wounded that escaped. After going through the pockets of the enemy casualties, Pittman handed Corporal Lovin a photo and ID card of the Chinese infantryman who had wounded him. Approximately two hours after the above action, Sergeant Pittman would be recognized with the Medal of Honor for throwing himself on a grenade which had landed in the midst of his squad.

Later that day, Regimental Commander Colonel Paul L. Freeman and Battalion Commander Lt. Col. Claire Hutchin Jr. visited the hill. Freeman was fluent in Chinese and interrogated some of the prisoners while Hutchin visited with the groggy Lovin. "Hell of a souvenir," Hutchin stated, looking at the newly air-conditioned helmet. Lovin asked if he could send it home, and Hutchin gladly signed a statement of contents which would accompany it.

Corporal Lovin wore a winter pile cap until captured three days later on November 30. He remained in North Korean prisons until he was repatriated on September 3, 1953. When he returned home he found his helmet safely stored at his grandmother's home where he had sent it a lifetime ago.

Interior view of the bullet's trajectory.

Left: The photo of the young Chinese soldier who had shot Corporal Lovin.

Below: It was a common practice for American soldiers to mark their belongings, including their helmets, with the first letter of their last name and the last four numerals of their serial number. In this case, "L" for Lovin and "1032."

Famous for uses other than protection, members of the 5th Marines give their puppy (called "Korea the 2nd") a much-needed bath. Korea, 5 September 1950. (US Army Photo, Courtesy of the Quartermaster Museum)

Intended to protect primarily from shrapnel, this photo graphically displays the ineffectiveness of the steel helmet against high-velocity rifle ammunition. Corporal Kenneth Fretty of Dexter Minnesota displays holes in his M1 helmet after a near miss. Fretty was injured by a bullet in the shoulder. (US Army Photo, Courtesy of the Quartermaster Museum)

General James A. Van Fleet in a bunker in Korea. He was commander of the U.S. 8th Army in Korea from April 1951 until his retirement on 31 March 1953. Note the multiple decals on his helmet.

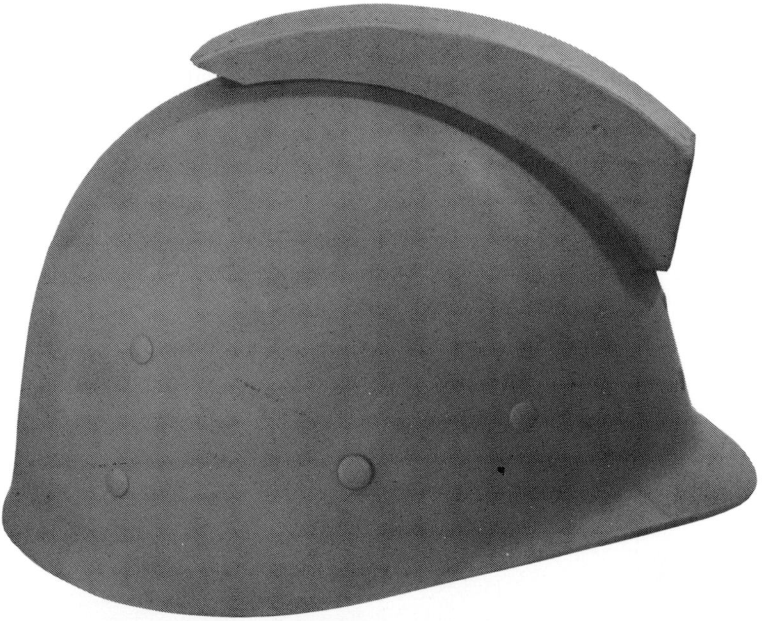

This is an example of a 1950s "Aggressor" or Opposition Forces (OPFOR) training helmet. These altered M1 helmet liners were used in the 1950s and 1960s in training exercises. They essentially identify friend from foe. The comb was constructed of wood and was held in place by screws, which caused the liner to be unserviceable for combat. In 1958 a clip system was developed so the combs could be attached to the liners without damage.

Second style OPFOR helmet used since the 1980s. It consists of a plastic "Soviet"-style helmet with a US helmet liner. (Courtesy of Col. Barry Smith)

All the internal liner components were installed and appeared essentially the same as the final version of the M1 helmet from WWII.

The first major modification to the M1 helmet liner appeared in the form of a new color for helmet liner webbing. Like its WWII cousin, the new webbing was made of cotton in a herringbone twill pattern. The main difference was the use of olive drab shade No. 7 rather than khaki. This darker color, accepted as the primary combat color for the U.S. Army in 1943, was used on all the internal suspension webbing inside the helmet. This included the neckband, headband and even the "A" frame suspension used on airborne configured liners.

A typical example of a Korean War manufactured liner. While it can't be confirmed that steel shells were made during the Korean War, there is no doubt that liners were. The primary difference between these liners and those made in WWII is the suspensions were a very dark OD on a herringbone twill weave. Most Korean War made liners are also dated in the crown of the liner.

Of the nine companies making helmet liners during WWII, most moved onto peacetime production of civilian products at the war's end. However, Westinghouse and several of its divisions, such as Micarta, along with the Capac Company, continued to manufacture helmet liners.

It appears the majority of liners made during the 1950s bore some type of date markings. Capac, for example, stamped the year of manufacture inside the crown of the liner with the maker's mark. Micarta tended to use a small sticker or adhesive patch with the year of manufacture.

Airborne liners were also manufactured during the Korean War. The primary difference between these liners and those made during WWII is the "A" frame webbing is installed outside of the basic helmet suspension with separate rivets. Like the standard liner, the airborne liner also used No. 7 herringbone-twill material.

The biggest change to helmet liners involved liners for airborne use. These liners, destined to be used with airborne configured shells, remained the primary domain of the Westinghouse Corporation.

For those interested in accurately determining if an airborne liner is from WWII or the Korean era, the key is the method used to secure the "A" frame webbing from the chinstrap to the liner shell.

During WWII, three manufacturers converted liners for airborne operations. These makers were Hawley,[4] Inland, and Westinghouse. In these WWII examples, the "A" frame webbing is placed beneath the existing suspension webbing and held in place by existing rivets.

Korean era examples were not manufactured using this method.

Rather, the "A" frame webbing is positioned below the liner suspension and independently riveted in place. The rivets in these 1950s version are larger than the standard rivets used to hold the suspension in place. Consequently,

[4] *While the shells were standard liners made by Hawley, McCord Radiator of Detroit modified them to the airborne configuration by partially disassembling them and adding the "A" frame chinstrap webbing. These were the first airborne helmet liners. Westinghouse modified their liners and the Inland made Liners. Some Capac liners are believed to have been modified by Westinghouse as well.*

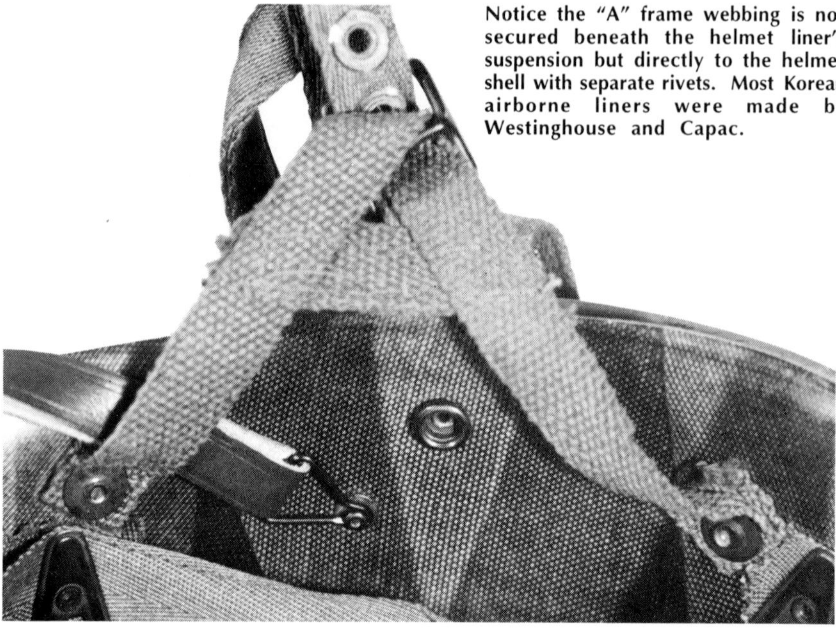

Notice the "A" frame webbing is not secured beneath the helmet liner's suspension but directly to the helmet shell with separate rivets. Most Korean airborne liners were made by Westinghouse and Capac.

the installation of the "A" frame chinstrap webbing accounted for an additional four rivets on the helmet liner shell.

Korean-era airborne liners retained the female snap. With the advent of the new T1 chinstrap clamp, an additional length of chinstrap webbing with a male fastener was used to secure the liner to the steel shell. Korean airborne configured shells can be found using T1 clamps and bar-tacked chinstraps on the steel shell.

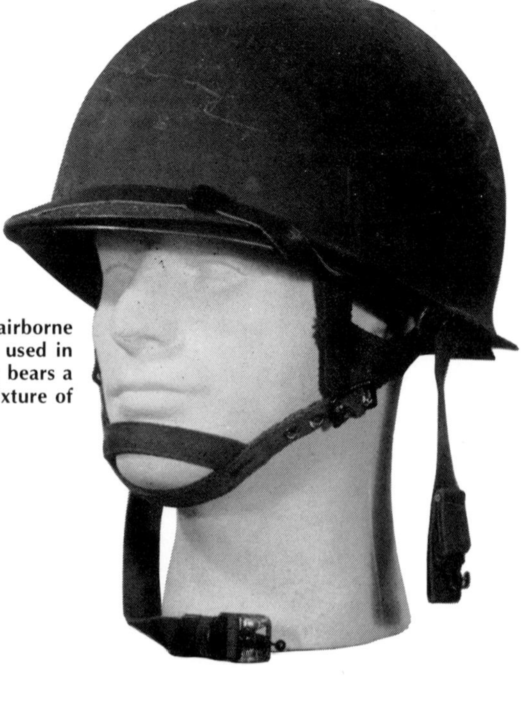

A typical Korean War helmet in airborne configuration. Like many helmets used in the Korean War this shell no longer bears a cork texture, but relies on a fine mixture of sand or silica gel.

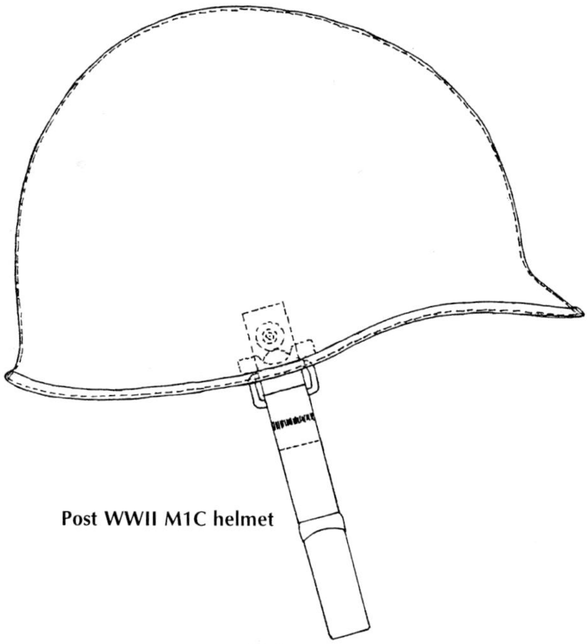

Post WWII M1C helmet

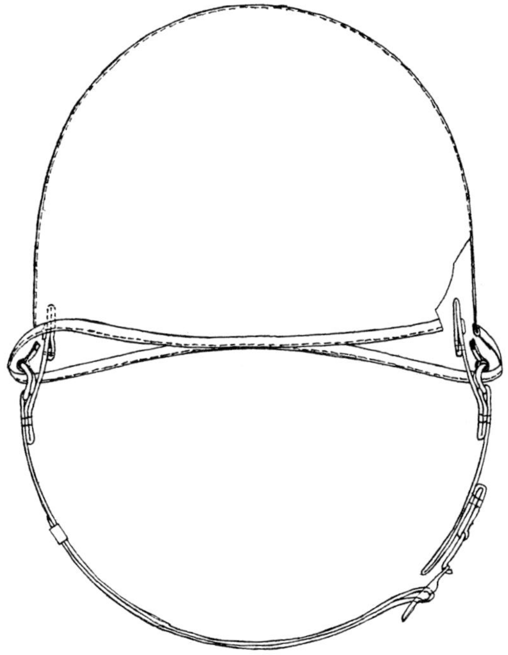

Like regular M1 liners, many of the Korean era airborne liners are marked and dated. Among collectors there is always talk of "Late WWII modified liners." Be aware that all things airborne are highly prized by collectors and unscrupulous individuals are known to be reproducing both M1C and M2 helmets, plus liners. Talk of late World War II modifications is speculation.

A member of the 55th Quartermaster Salvage Depot in Korea repaints M1 helmet liners, 1 June 1951. This practice was not limited to liners. Steel shells also were repaired and refinished. Field conditions were not ideal for color matching or consistency in finish. Consequently M1 steel helmets and liners can be encountered in a variety of OD shades. It is interesting to note this soldier is wearing a one-piece HBT suit with USAF Staff Sergeant stripes removed. (US Army Photo, Courtesy of the Quartermaster Museum)

An interesting perspective of the variations in color one can expect to encounter with post-WWII M1 helmets. Both of these are in the Korean War configuration but are painted with different shades of OD. Due to being cleaned in a phosphate solution prior to painting, the cork finish applied when the shells were made in the 1940s dissolved.

During the 1950s both the U.S. manufactured helmet net used during WWII (complete with neoprene band), and the U.S. Marine Corps camouflage helmet cover continued to be produced. Unlike during WWII, the USMC helmet cover was now produced in only one version, and was normally marked and dated on the brown side.

The USMC continued its practice of using camouflage helmet covers. Most post WWII covers are identical to the WWII 2nd pattern cover but are usually marked and dated on the "brown side." (The Larry Munnikhuysen III Collection)

A variant of a postwar USMC cover. Interestingly, this cover does not have foliage slits and the detailed EGA insignia was printed on a separate diamond-shaped piece of camouflage material and sewn in place. (Courtesy of John Coy)

A USMC helmet cover dated 1953, as worn on the M1 helmet. 1953 is the only contract date that has been observed. (The Larry Munnikhuysen III Collection)

The interesting cover and cape system displayed in these photos was probably designed to protect the helmet and shoulders from minute particles of nuclear fallout. The cover could also be used as a sunshield in desert operating theaters.

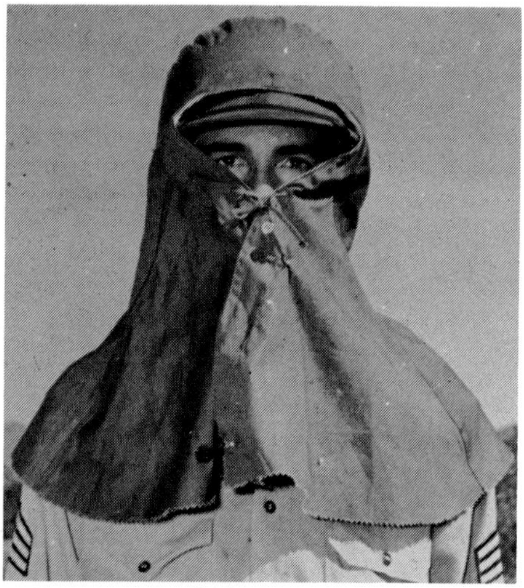

Some time in 1958, a batch of M1 helmet shells was produced. Unfortunately the exact number pressed and the name of the manufacturer was not able to be determined.[5]

The late 1950s brought the biggest change to date to the M1 helmet and its liner, as approval was given to convert the material used in the liner from duckcloth to laminated nylon. The change to nylon became official in 1958, although with millions of duckcloth liners still in use, it would be the middle 1960's before the nylon liners would be in the majority.

The material used in the suspension also changed. It appears 1958 was the year the helmet liner's OD Herringbone Twill suspension was exchanged for a sturdier, thicker, cotton webbing. This new webbing was slightly duller in color than standard olive drab No. 7. The internal suspension continued, however, to be installed in the same manner as the WWII and Korean era liners.

A comparison of headband clips and material. Top, the No. 7 color used in the majority of Korean War era helmets. Bottom: The new weave was introduced in 1958.

Although they would continue to be visible for years due to existing stocks, It appears that 1958 signaled the end of the insignia eyelet rivet at the front of the liner and the leather chinstrap and studs.

---

[5] *A newspaper clipping from The New York Times dated 1965 states that the U.S. Army pressed nearly one million new helmet shells in that year. The clipping remarked that it was the first new batch pressed since 1958.*

In 1958 the primary material used in M-1 helmet liners changed from duckcloth to nylon. The insignia eyelet became a thing of the past and was eliminated in approximately 1958.

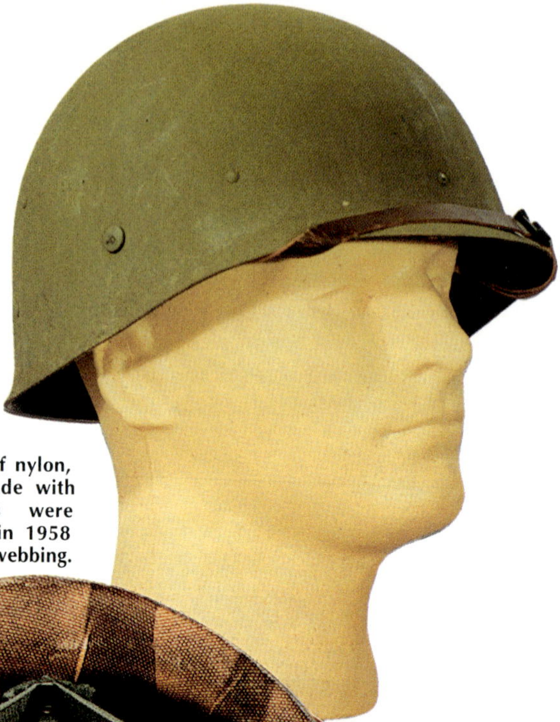

Below: Even with the advent of nylon, most liners would still be made with duckcloth until supplies were exhausted. This liner, made in 1958 displays the new color cotton webbing.

Constructed of a heavy composite material similar to a liner, this helmet, designated the AN/PRC 34 (X1) was created in the late 1950s. It used a paratrooper style chinstrap and sported a strange looking antenna. It held a self-contained two-way radio system. There was no steel shell for this helmet.

The AN/PRC 34 (X1) held a variety of items including batteries, a transmitter, a receiver, microphone, and ear phones. They were made by RCA and several thousand were manufactured. They were used in the late 1950s and into the Vietnam war. This example is missing many of the internal radio components. The helmet was very heavy.

Developed in the late 1950s, the AN-PRC 36 (X1) was an attempt to provide improved radio communication capabilities to the combat infantryman. The system was powered by a battery enclosed in a case, clipped onto the web belt. A cable ran from the belt to the antenna and speaker, which was attached to the M1 helmet shell on an elastic band held in place by three small metal clips.

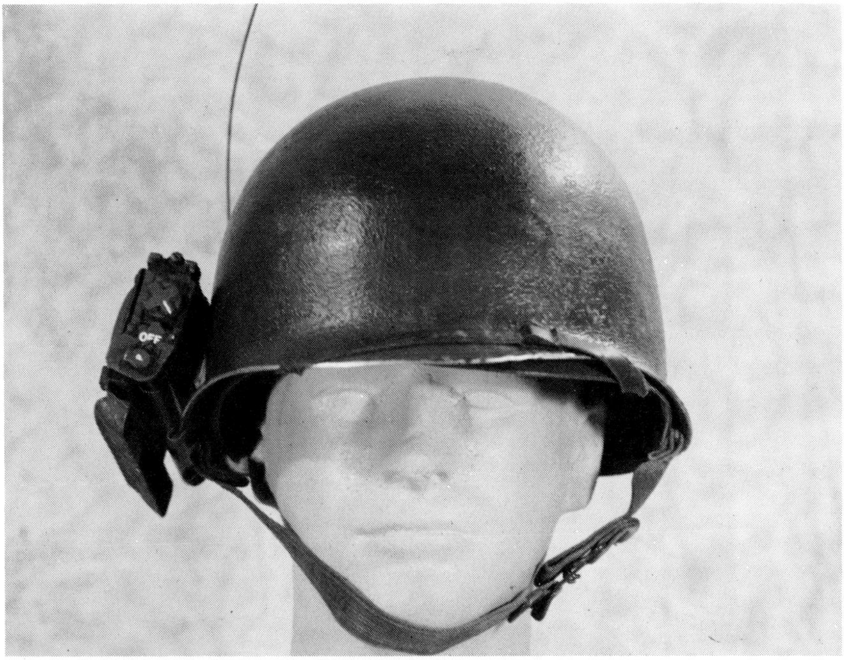

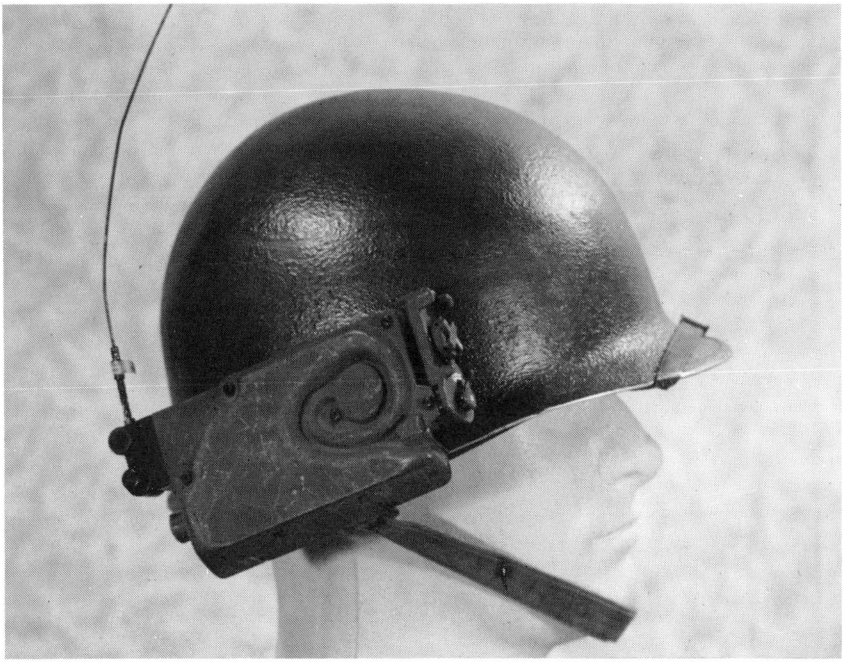

Another variation of the helmet radio, the AN-PRC 9 clip-on helmet radio receiver, which was clipped to the side of a standard M1 helmet shell. The AN-PRC 9 was battery-powered and saw service during the Vietnam war. This example was dated 1969.

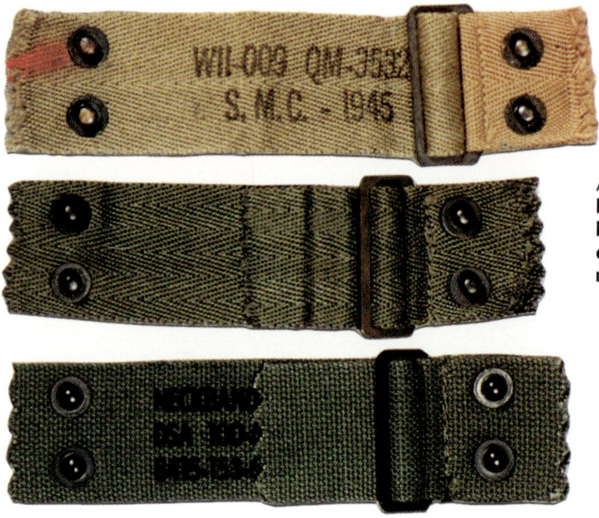

A comparison of neck bands. From top to bottom: WWII, Korean era, and the new 1958 material and color.

In 1963 the ERDL pattern helmet cover was adopted as the standard cover for all branches of service. Identical in construction to the USMC covers of the 1940s and 50s, the new cover featured a design created in the 1940s by the Army "Environmental Research and Development Laboratory" (ERDL).

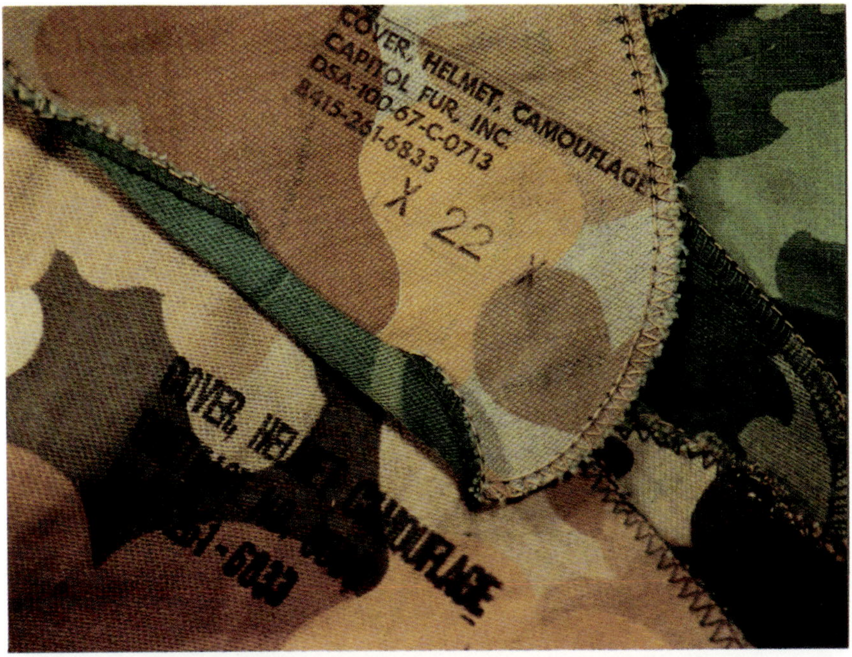

The vast majority of the ERDL issue helmet covers are marked and dated on the "brown" side. There was an assortment of companies who made the cover.

The Vietnam Era

During the Vietnam war, the U.S. military continued to rely on the venerable M1 helmet, which proved to be a recyclable resource for the Army. Many veterans confirm being issued not only WWII vintage helmets and liners in the mid 1960s, but web gear as well.[6]

As during the 1950s, alterations to the steel shell remained minimal. In the early 1960s cork was eliminated from the texture. A fine sand aggregate would remain as the single texturing ingredient for the helmet. Beyond that, the helmet shell remained essentially unchanged.

The liner, however, continued its metamorphosis. In 1965 the laminated nylon liner was now in full production. Consequently, the classic dark brown duckcloth interior was replaced by green, orange, or a mixture of the two. The suspension also underwent significant modifications. In 1965 an additional 1,000,000 M1 steel helmet shells were manufactured by Ingersol Products Division at a price of $3.60 each.

The original suspension used in WWII and Korea relied on three lengths of webbing tied with a shoestring into a hammock at the crown of the liner. The ends of the webbing lengths were permanently riveted in place. An additional length of webbing circled the interior of the liner. When installed the headband was clipped to this piece of webbing. The final length of webbing was riveted to the rear of the helmet to support the neckband.

---

[6] *From an interview with USMC vet Don Leffel. This continued into the 1980s. During a tour of duty in Europe in 1984, the author was issued a Vietnam era steel shell with a repainted WWII Capac manufactured khaki suspension liner!*

During Operation Back Pack in Southern Taiwan, Lt Philbrook S. Cushing, battalion landing team talks to the air support. The helmet cover holding the twigs is the ERDL camouflage model.

Staff Sergeant Frank A. Martin, 1st Battalion, 7th Artillery, 1st Infantry Division during a stand-down at Sai Khe base camp, Vietnam 1968. Staff Sergeant Martin's helmet sports the ERDL helmet cover.

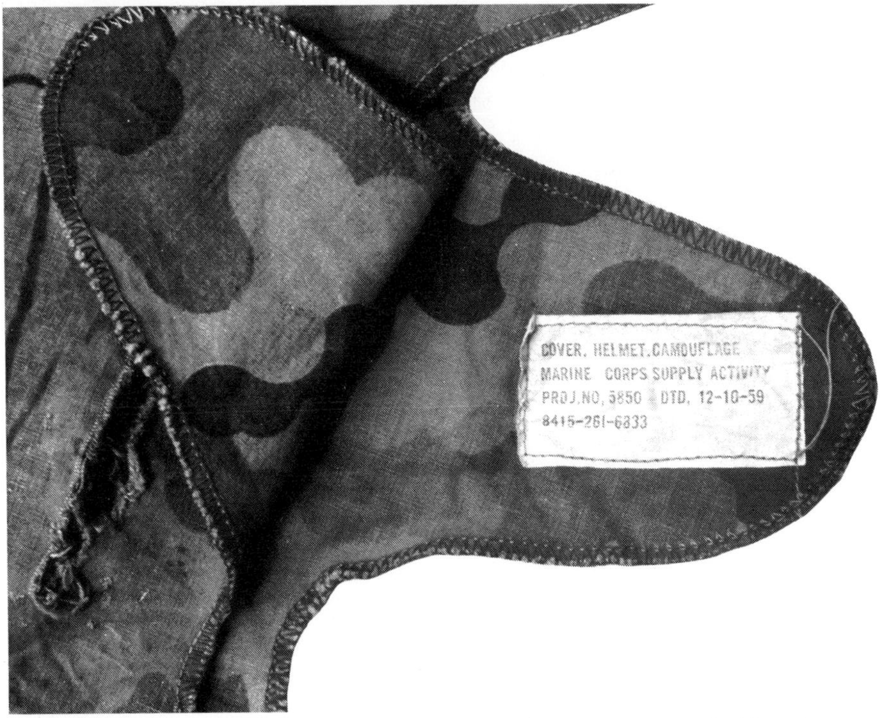

The new helmet cover was developed in the 40s, but did not become a standard issue item until approximately 1963. Examples with earlier dates, however, are sometimes encountered. Many equipment items went through years of development and testing prior to full adoption for use. This 1959 dated cover may be one of these early evaluation items.

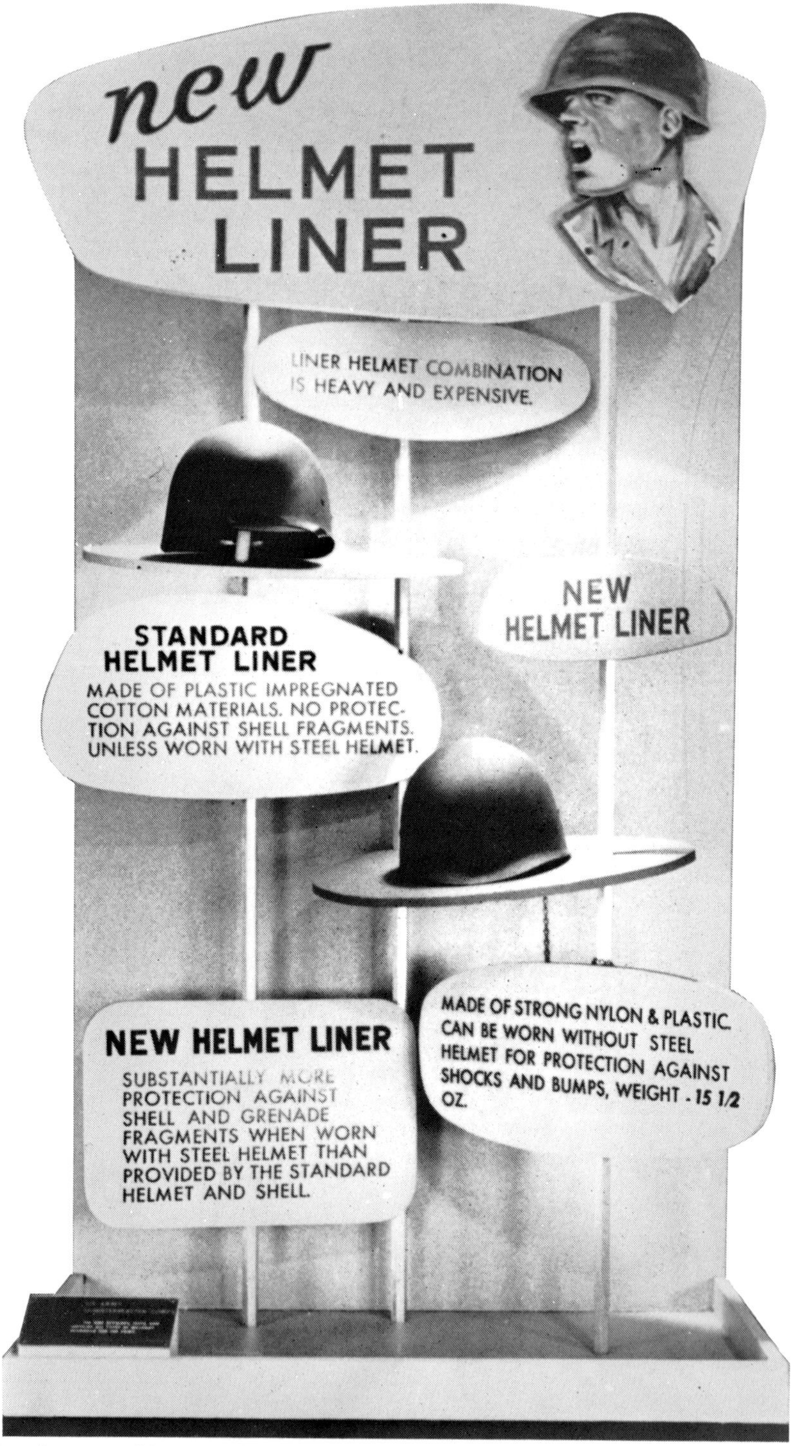

Although approved in 1958, the use of Nylon did not become widespread in helmet liners until the middle 1960s. This is an Army display touting the virtues of the new nylon liner. (U.S. Army Photo, Courtesy of the Quartermaster Museum)

In 1961 the nylon helmet liners started to appear. The materials used resulted in interesting orange and greens inside the liner shell. Notice the new configuration of the suspension. The hammock style and shoestring cord were eliminated in the late 1960s.

If the webbing inside the WWII or Korean liner was damaged or cut, the liner was normally discarded because there was no specified method to repair those types of defects.

The suspension of 1965 was very different.

Using the thicker 1958 webbing material, the liner suspension, like that in the WWII or Korean-era helmets, circled the interior of the liner shell. It could, however, be removed and replaced. Instead of permanently riveting the suspension to the shell, clips, which resembled washers, were riveted inside the shell. The mates to these washer-type clips were riveted to the web suspension. The suspension could simply be clipped in place and could be removed for cleaning or replacement.

The configuration of the webbing changed as well. Instead of the hammock method, three lengths of webbing crossed each other at the peak of the interior of the liner. Each could be adjusted at one end with a buckle identical to the buckle on the WWII or Korean-era helmet liner neck strap.

Speaking of neck straps, that item completely changed with the 1965 liner. The suspension webbing which supported the neck strap in the WWII

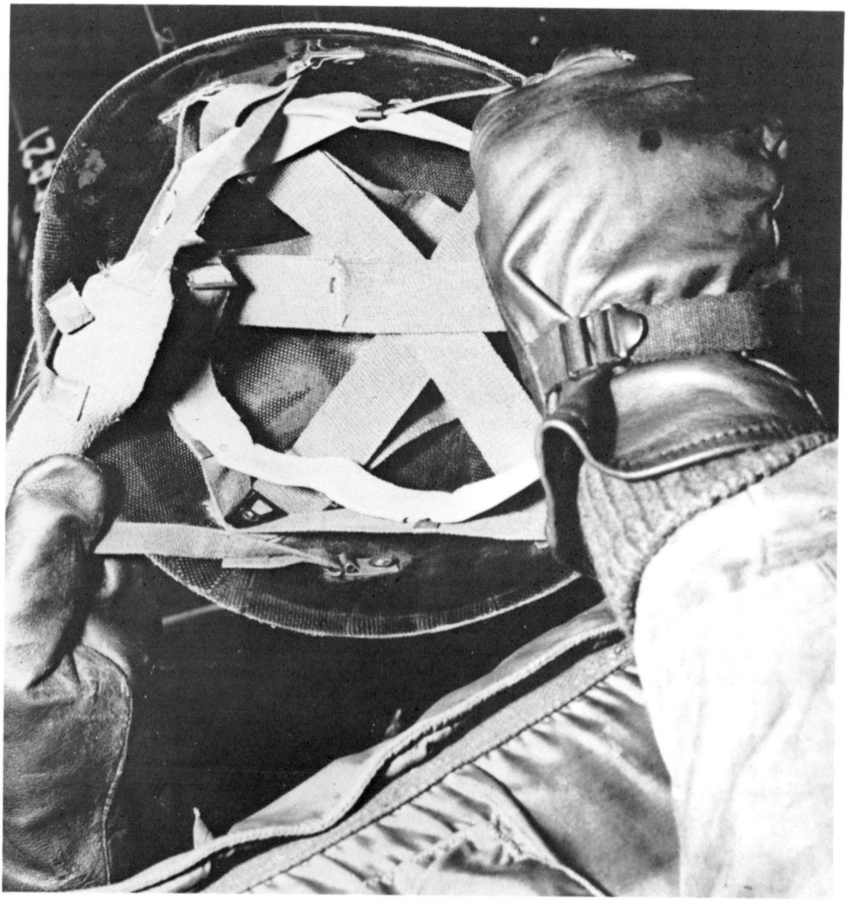

A U.S. Army photo showing the new neckband. Notice how the band is held in place by a buckle system. (US Army Photo, Courtesy of the Quartermaster Museum)

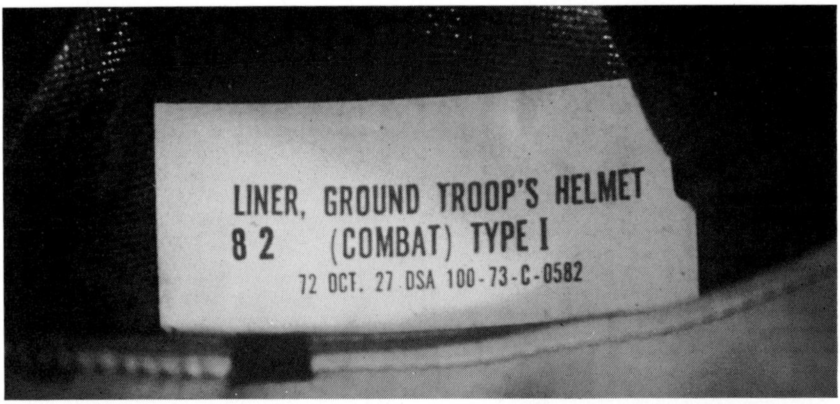

A paper nomenclature tag describing the helmet. This example is from 1972.

and Korean liner was no longer was installed. Taking its place were three clips, riveted along the interior of the liner: One at the rear of the helmet and one on each side of the helmet.

A new style of neckband, which gave support to the back of the head, was laced through these clips. Like the new suspension, the new neckband could be removed and replaced.

The exterior of the helmet liner was painted and textured in a color slightly darker than the steel shell. Liners were now designated as "Ground Troops" or "Parachutists."

The parachutist's liner was constructed identically to the ground troop's version with the exception of the "A" frame webbing used to support the airborne chinstrap. While the suspension and neckband could be removed and

Airborne helmet liners from the Vietnam era were made in the same manner as those from the Korean era with the "A" frame chinstrap supports attached separately from the suspension.

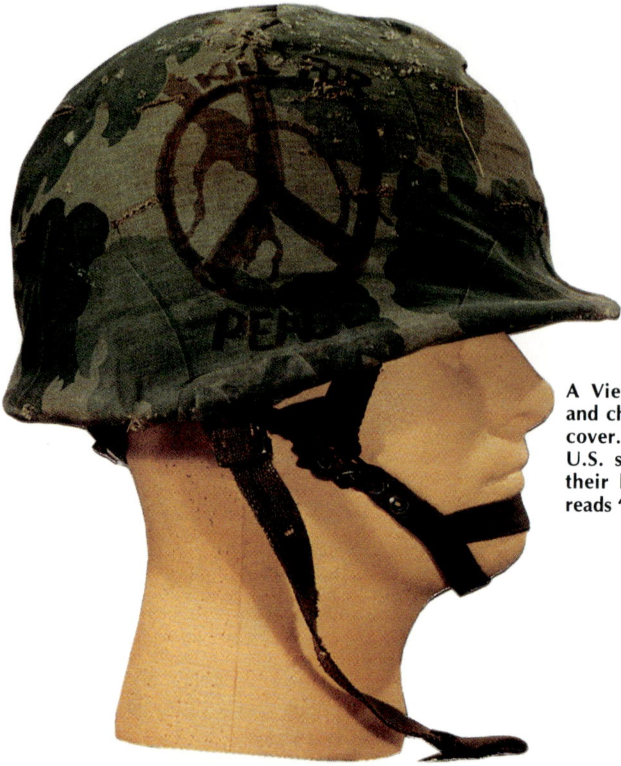

A Vietnam-era airborne helmet and chinstrap with ERDL helmet cover. During the Vietnam era, U.S. soldiers often personalized their helmet covers. This one reads "Kill for Peace."

A holdover from W.W.II, the neoprene band continued to serve as a method to hold the helmet cover securely to the helmet shell. The neoprene band is still in use today.

In Vietnam, U.S. forces often tucked items such as cigarettes and insect repellent beneath the neoprene band. (Courtesy of Glenn Haley)

This platoon leader wears a M1 helmet with ERDL helmet cover and neoprene band in the 1970s. (US Army via George A. Petersen)

replaced from the airborne helmet liner, the "A" frame webbing, however, was riveted directly to the liner shell and could not be removed.

While the USMC still retained its WWII and Korean pattern helmet cover, the Army drifted away from the use of the camouflage net.

In approximately 1963, the Army's Environmental Research and Development Laboratories (ERDL) patterns of camouflage were adopted for use in clothing. Interestingly, the Army adopted a camouflaged helmet cover[7] which did not match the ERDL developed uniform designs.

The cover was a 100 percent cotton cloth, printed on two sides and had 16 slits for inserting foliage. Like USMC covers of WWII and Korea, the camouflaged cover had a "green" side and a "brown" side and was identical in the method of construction. While the patterns on both sides of the cover simulate leaves and twigs, strangely these "leaves and twigs" resemble foliage found in a forest in Wisconsin or Ohio rather than a jungle in Southeast Asia.

Worn primarily with OD jungle fatigues and often held in place by a neoprene helmet band similar to the ones used on the obsolete helmet net, the cover replaced even the Marine Corps old stand-by.

The majority of the covers are maker marked, with the marks and stock numbers usually located on one of the "brown side" leaves. There were a variety of manufacturers.

The 1970s and 1980s

As the 60s changed to the 70s, the M1 helmet and its liner also continued to evolve. In 1972, the M1 helmet chinstrap changed. The new webbed chinstrap incorporated the paratroopers style cupped chinstrap. This, coupled with a pull-the-snap release similar to that used on the WWII M3 anti-flak helmet, helped hold the helmet more securely. In approximately 1975 the

An assortment of post-W.W.II M-1 helmet chinstraps. Top: The final M-1 helmet chinstrap from the early 1980s, Center: The standard chinstrap used from the Korean War through the 1970s, Bottom: The airborne style chinstrap used from 1944 through the 1980s.

clamp on T1 chinstraps was replaced by a new chinstrap clip. This clip was snapped over the chinstrap loop on the M1 shell. Although the clips fit tight and held securely, they could swing freely on the chinstrap loop. The advantage the new clip was ease of removal and replacement. With the helmet off, a sharp twist to the rear on the chinstrap removed the clip from the loop.

Now, instead of struggling to twist and tug to unlatch the old hook and ball mechanism of the T1 clip, a soldier just needed a sharp tug to unsnap the strap.

Adopted in the 1980s the helmet chinstrap changed to nylon. This clip was a new design which slipped over the helmet chinstrap loop. It also used a cam buckle system to allow the strap to be adjusted.

The final configuration of the M1 steel helmet. This photo is representative of helmets used from the middle 1970s to the 1990s. The new helmet relied on the new nylon chinstrap with a press-the-dot release.

The Army's field manual 21-15, Care and Use of Individual Clothing and Equipment, advises GI's on the proper procedures for removing the old style T1 clamps from their helmets.

It appears however, that paratrooper helmets retained the old-style T1 clamps and chinstrap with the additional tab and snap.

Another product introduced in approximately 1975 was the woodland camouflage helmet cover.

[7] *Author's note: It appears the new cover saw widespread issue starting in 1963. However, earlier dated examples are sometimes encountered.*

Unlike the Vietnam camouflage pattern which failed to match ERDL uniform patterns, the new woodland camouflage, which used a mixture of brown, black, khaki and primarily green, matched the Army and Marine Corps new camouflaged battle dress uniform perfectly.

As with previous covers, the woodland cover was made of cotton and had slits to add natural camouflage by inserting foliage. It was, however, a single-sided cover. Gone was the reversible "sand brown" side. The interior of the woodland camo cover is a pale green. Like the Vietnam ERDL cover, maker's marks and stock-number data can usually be found on one of the leaves.

The 1980s saw the final production liners for the M1. These final versions appeared more like plastic than any previous. The external paint color was pale green.

Introduced in approximately 1973, the woodland camouflage pattern replaced the ERDL camouflage cover. Unlike earlier covers this new cover was not reversible.

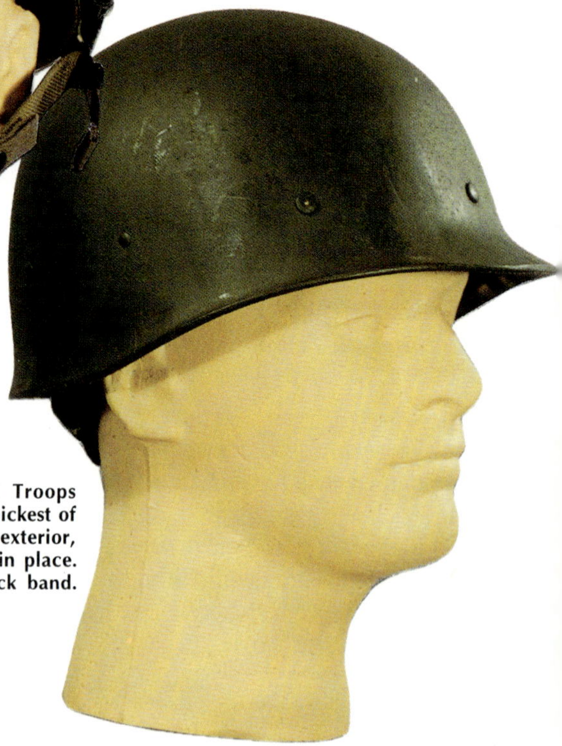

The final version of the Ground Troops Helmet Liner, Type One. It is the thickest of all the plastic or nylon liners. On the exterior, six large rivets hold the suspension in place. Three smaller rivets support the neck band.

A peek inside the final ground version of the M1 helmet liner as manufactured in the 1980s.

As with the 1970s liners these final versions permitted removal of the headband. Six slip-on clips were riveted to the headband. These were matched with flat-washer style clips attached to the liner shell. The tension held the headband in place. This system was also used on the airborne helmet liner.

Typical markings found inside a late 1980s liner. Note the buckle for holding the neckband in place and the washer for the suspension.

While the ground troops helmet liner hardly changed in its final version, the final airborne liner added one last accouterment before passing into history.

A special foam rubber pad was made which could be inserted between the headband and neckband at the rear of the helmet. The obvious intent was to provide more support and reduce shock to the neck and head.

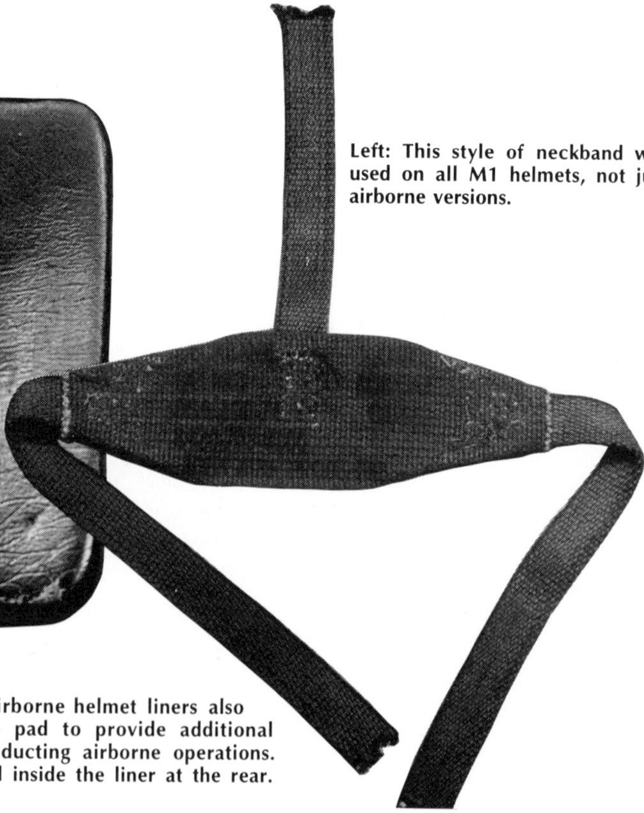
Left: This style of neckband was used on all M1 helmets, not just airborne versions.

By the 1980s most airborne helmet liners also included an Ensolite pad to provide additional protection when conducting airborne operations. The pad was installed inside the liner at the rear.

The inside of a 1980s-made M1 airborne liner. Note the Ensolite pad installed at the rear of the liner.

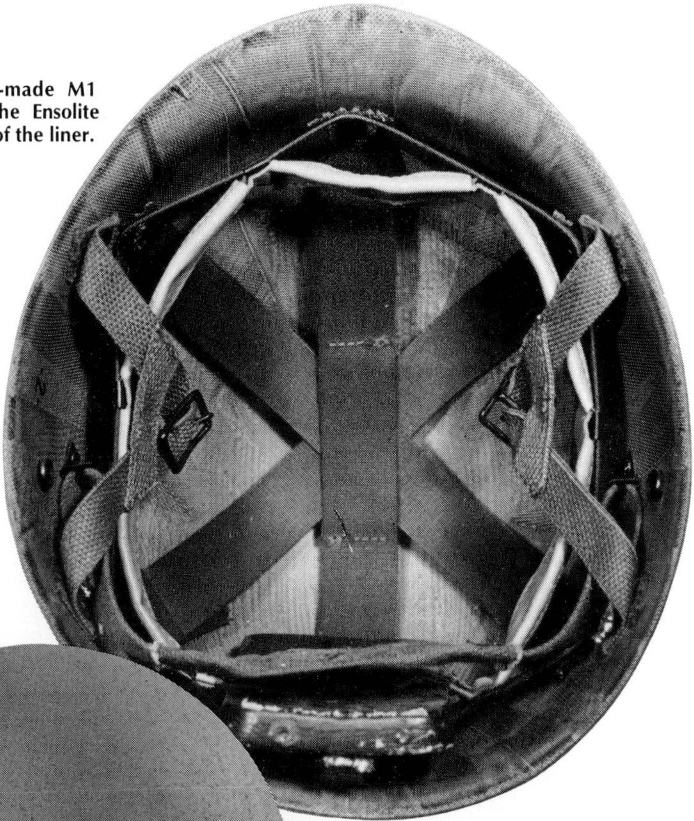

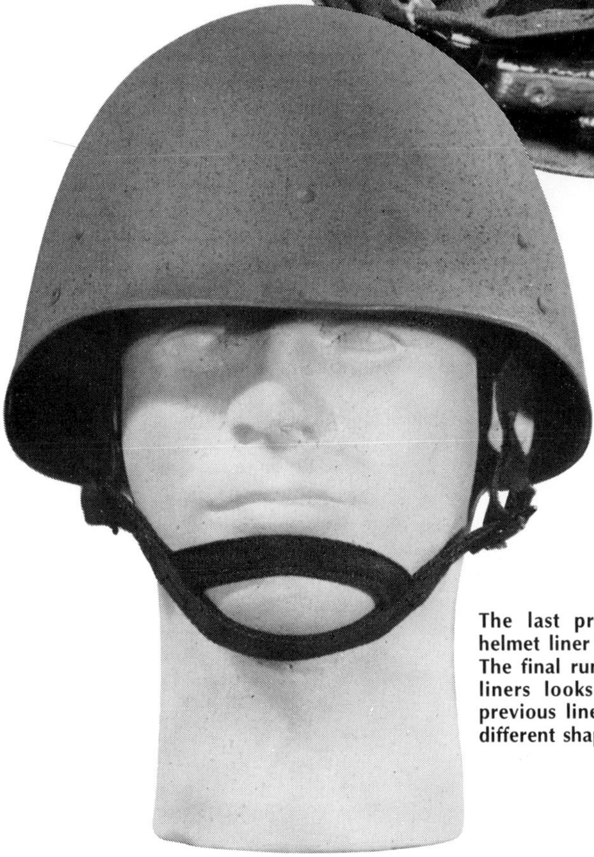

The last production version of the M1 helmet liner in the airborne configuration. The final run of both ground and airborne liners looks more like plastic than any previous liners. They also have a slightly different shape.

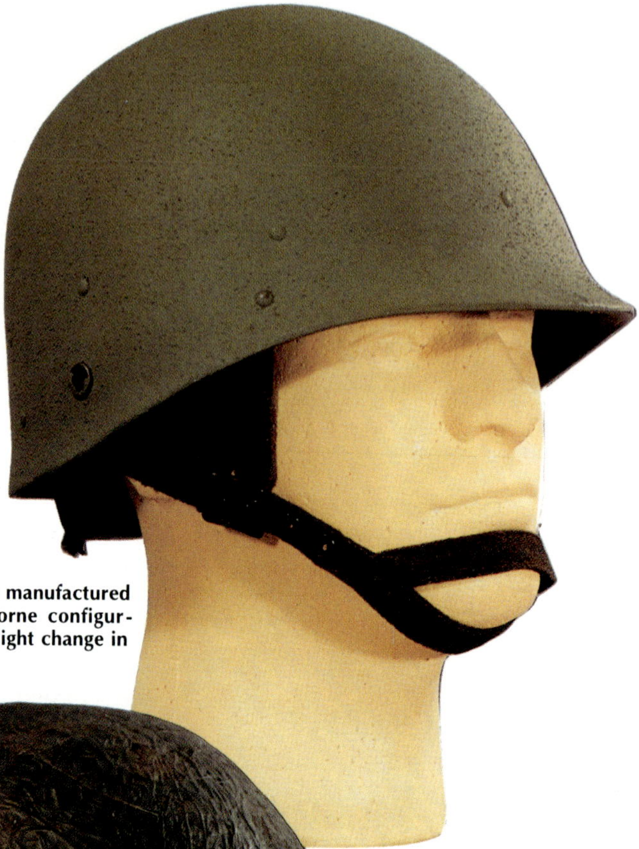

A final view of the 1980s manufactured M1 helmet liner in airborne configuration Again, notice the slight change in overall appearance.

Made from an epoxy, plastic helmet shells such as this were reported to have been made near Fort Bragg and other Army installations for instructors. These very light shells fit over M1 liners, and with a helmet cover in place an instructor could avoid wearing the full three pound steel pot!

An experimental M1 helmet developed during the Vietnam era. The helmet uses a four-point strap retention system. Riveted and welded to the underside of the steel helmet shell are four swiveling chinstrap loops. Through these loops pass the nylon straps. The cradle for the chin is made from cotton. This helmet uses a standard M1 helmet liner as its suspension system. (Courtesy of Bill Rizzutti)

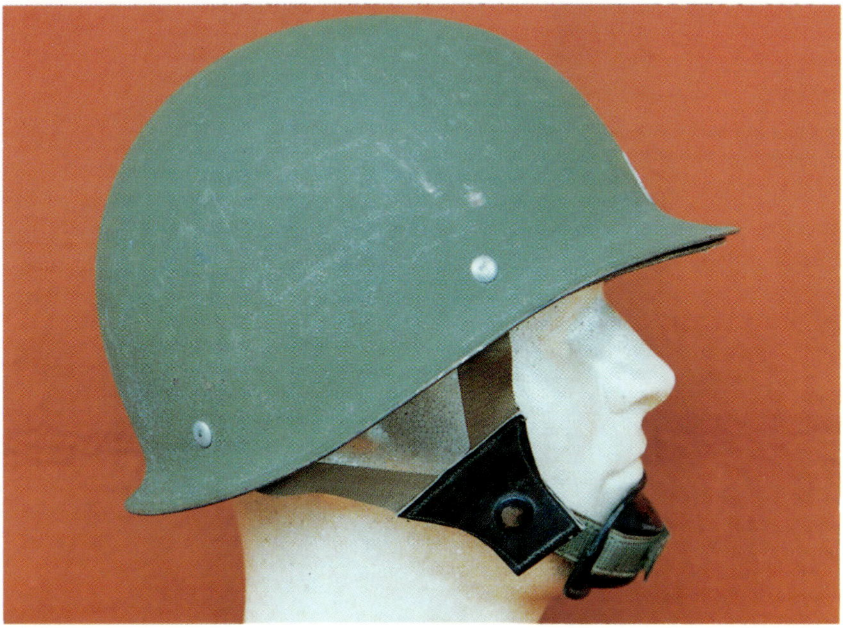

Another Vietnam era experimental chinstrap configuration which used a plastic chin cup rather than a cotton chin cradle. The anchors for the four-point chinstrap are riveted to the side of a standard M1 helmet shell. This helmet also uses a standard M1 liner. (Courtesy of Bill Rizzutti)

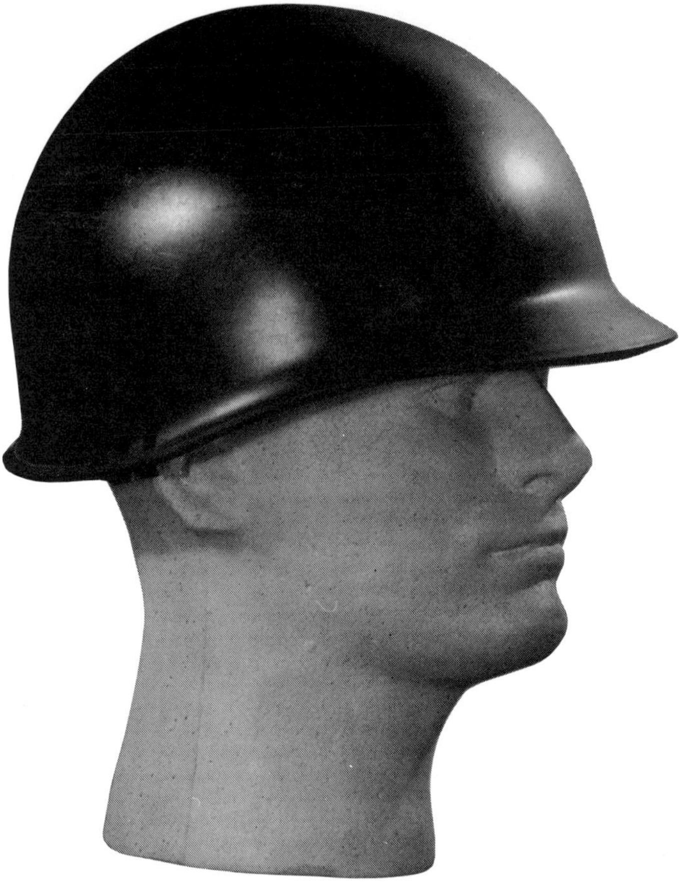

Another interesting M1 shell variant made from a very hard, high-impact plastic. The purpose of this helmet and when it was made are unknown.

In 1983, during the invasion of Grenada, the U.S. military unveiled the successor to the venerable M1. Although most regular Army and Marine Corps units have replaced the M1, they can still be found in service at many Air Force and Navy bases, not to mention Guard and Reserve units of all services.

A front view of these unusual visored helmets. (Courtesy of Bill Rizzutti)

Designed to be worn over a standard M1 steel helmet, the exact purpose of these protective, visored helmets is not known. Possibly they were designed for mine clearing or bomb disposal purposes. The helmets and visors are made from a non-metallic material. The viewing ports are a Plexiglass material secured to the helmet with screws. (Courtesy of Bill Rizzutti)

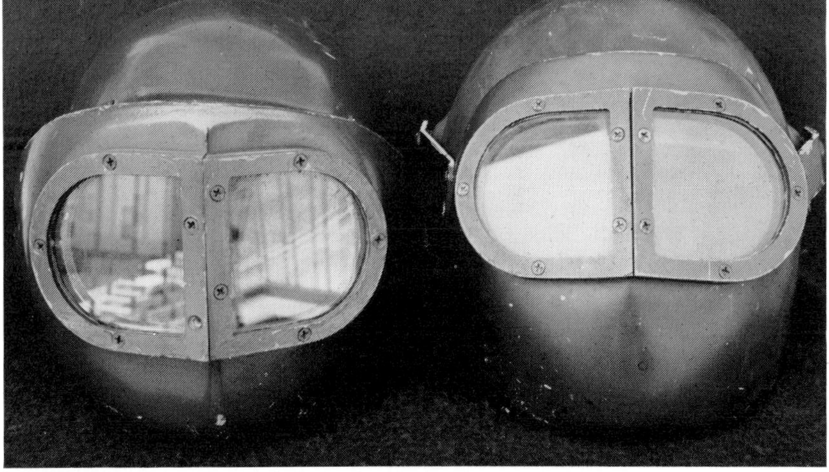

CHAPTER SEVEN

STRONGER THAN STEEL: THE PASGT BALLISTIC HELMET

In the 1950s Army researchers and medical personnel combined forces to analyze the protective capabilities of the M1 helmet during World War II and Korea. Among the efforts was special research project, number 81-01-09.1, Military Helmet Design, compiled at Camp Lejune, NC. This project reviewed the history, design, development, production and use of helmets from around the world. It also attempted to address medical, psychological, logistical and tactical aspects of helmet development. This report, plus others, provided a series of recommendations to the U.S. military which would start the long process to improve the American soldier's No. 1 piece of protective armor, the helmet. Among the recommendations were:

- Replace the one-size-fits-all steel shell of the M1 with a helmet available in three sizes.
- Provide some protection against radiation, heat, wind and cold.
- Consider ballistic materials other than steel.

From approximately 1948 through 1955, the U.S. Army conducted research to develop a new combat helmet. During the experimentation stage, a variety of materials and designs would be tested. A prime focus of the new helmet study was to develop a helmet which could be used by an assortment of forces: ground, armored, anti-aircraft, communications troops and even airborne elements. The first of the post WWII experimental helmets to be tested was the X-49 series. The X-49 helmet was similar to the M1 in that it was a two-piece helmet consisting of a shell and liner which held the suspension. It was a larger helmet and covered more of the head and sides. Like many past U.S. experimental helmets, the initial X-49 generation shells were designed and manufactured by hand at the Metropolitan Museum in New York. These two initial shells were made from sheet aluminum of different thickness and were designated X-49-1 (.051" sheet aluminum) and X-49-2 (.082" sheet aluminum). An unknown number of examples of both the X-49-1 and X-49-2 were then produced and used for testing with three new types of helmet liners made from an assortment of laminates and man-made compounds and nylons. The tests of the X-49 helmets were conducted at Ft. Benning, Georgia, using 33 soldiers who engaged in a variety of activities in simulated combat conditions over a two-week period. The test results revealed that the test partici-

During the 1950s, scientists were able to start assessing wound statistics kept during World War II. While the information confirmed the M1 helmet saved lives, it was determined the helmet could be improved. Thus began a 25-year journey toward the ballistic helmet. This is a 1950s era helmet. This helmet, designed in the early 1950s was for tropical use. It appears the material used in the construction is something other than steel. (U.S. Army Photo Courtesy of the Quartermaster Museum)

pants preferred the lighter X-49-1 helmet to the heavier X-49-2 model. While the helmets seemed to pass the test for infantry use, the X-49 helmets didn't fair as well with airborne, crew-served weapons, and tank forces. The chinstrap on the X-49 helmets was a larger version of the M1 helmet's T1 clip, but didn't work as well. It was difficult to use and in airborne tests, every paratrooper who jumped lost his X-49 helmet. Later X-49 helmets used helmet liners with "A" frame webbing chinstraps as used in the M1 jump helmets.

This is a six-ply nylon liner for an Experimental X-49-1 helmet. The suspension system is essentially identical to the M1 helmet liner. (Courtesy of Bill Rizzutti)

Armored forces found the X-49 helmet extremely difficult to use and that it did not provide adequate "bump" protection in comparison with the old WWII football-style helmet. Further testing required approximately 2,000 more X-49-1 helmets to be manufactured in two different sizes. The fragility of the helmet also made it a poor candidate to replace the M1. Obviously, the aluminum helmet shell was not designed to protect a soldier from direct bullet strikes, but coupled with its ballistic nylon liner, was intended to provide protection primarily from shrapnel. In early 1950, a third version of the X-49, the X-49-3 was tested. The X-49-3 was actually a liner constructed of a new, lighter type of nylon and not another version of the aluminum shell. Several

This is another variant of the Experimental X-49-1 liner. The main difference between this liner and the previous example is this liner is a five-ply-nylon liner. The suspension system is held in place using a steel band riveted to the inside of the liner shell. (Courtesy of Bill Rizzutti)

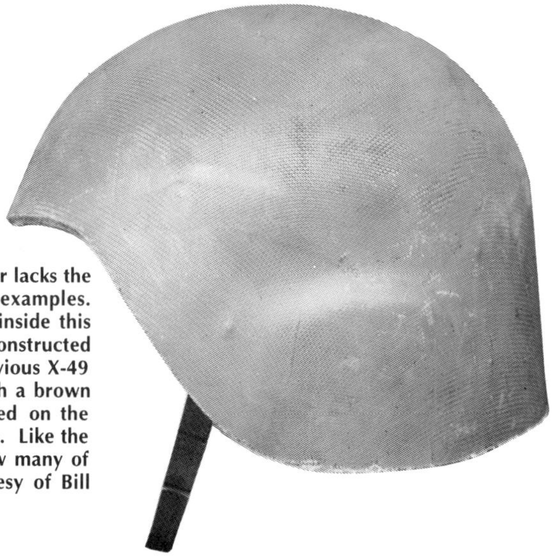

This Experimental X-49-1 helmet liner lacks the rivet system found on the previous examples. The suspension was held in place inside this liner using screws. This liner was constructed of seven-ply nylon. Unlike the previous X-49 liners, this example is equipped with a brown leather chinstrap. The washers used on the suspension are constructed of plastic. Like the other X-49 liners, it is unknown how many of these examples were made. (Courtesy of Bill Rizzutti)

different suspensions were used in the liner which confirmed that future helmets and liners needed to be designed in multiple sizes rather than the one-size-fits-all approach used with the M1. In the end, the X-49 helmet and its ballistic liner was rejected by the U.S. military.

Nylon was not the only material used in these early experimental attempts to replace duckcloth and steel. This X-49 helmet liner is made from a composite material called DORON. DORON was far more stiff and rigid than nylon. The liner retains the standard hammock configuration used in the M1 helmet liner. The suspension is held in place by using screws. (Courtesy of Bill Rizzutti)

The next attempt was the X-51 helmet. Looking very similar to the X-49, the X-51 shell was made of heat-treated aluminum. The X-51 chinstrap loops also doubled as locking mechanisms to hold the nylon X-51 liner to the aluminum shell. It was made in two sizes, both of which weighed less than one and one-half pounds. While it was hoped the X-51 could be used by a variety of forces, designers determined that armored crew users would only use the nylon liner system. The X-51 liner was the primary armor for the system and was made from a nine-ply basket weave of nylon. It had the same shape and contours as the shell and fit snugly inside it. Liners were also made in two sizes. In ballistic tests, the X-51 helmet and liner provided 30% greater armor protection against shrapnel than the M1. However, extended field test-

A very interesting look at an Experimental X-49 helmet liner in an airborne configuration. This helmet is equipped with a leather fleece-lined chin cup. It is very interesting to note that the "A" frame chinstrap webbing is attached beneath the standard suspension system in a manner very similar to WWII airborne M1 liners. Keep in mind this helmet was made in the late 1940s and the very dark OD number 7 herringbone twill webbing had not superseded the khaki webbing. (Courtesy of Bill Rizzutti)

ing declared the helmet unsuitable for use by nearly every type of combat position. The primary reasons were:

- The nylon liner tended to warp
- The helmet left a large part of the front of the head exposed
- The helmet parts often separated in use
- The helmet did not provide adequate protection against small arms fire
- The helmet hardware often broke or bent

Like the X-49 series, the X-51 helmet project was abandoned. Other experimental models such as the Doron helmet, the T21E1, X-53 and the T-54-1 suffered similar fates.

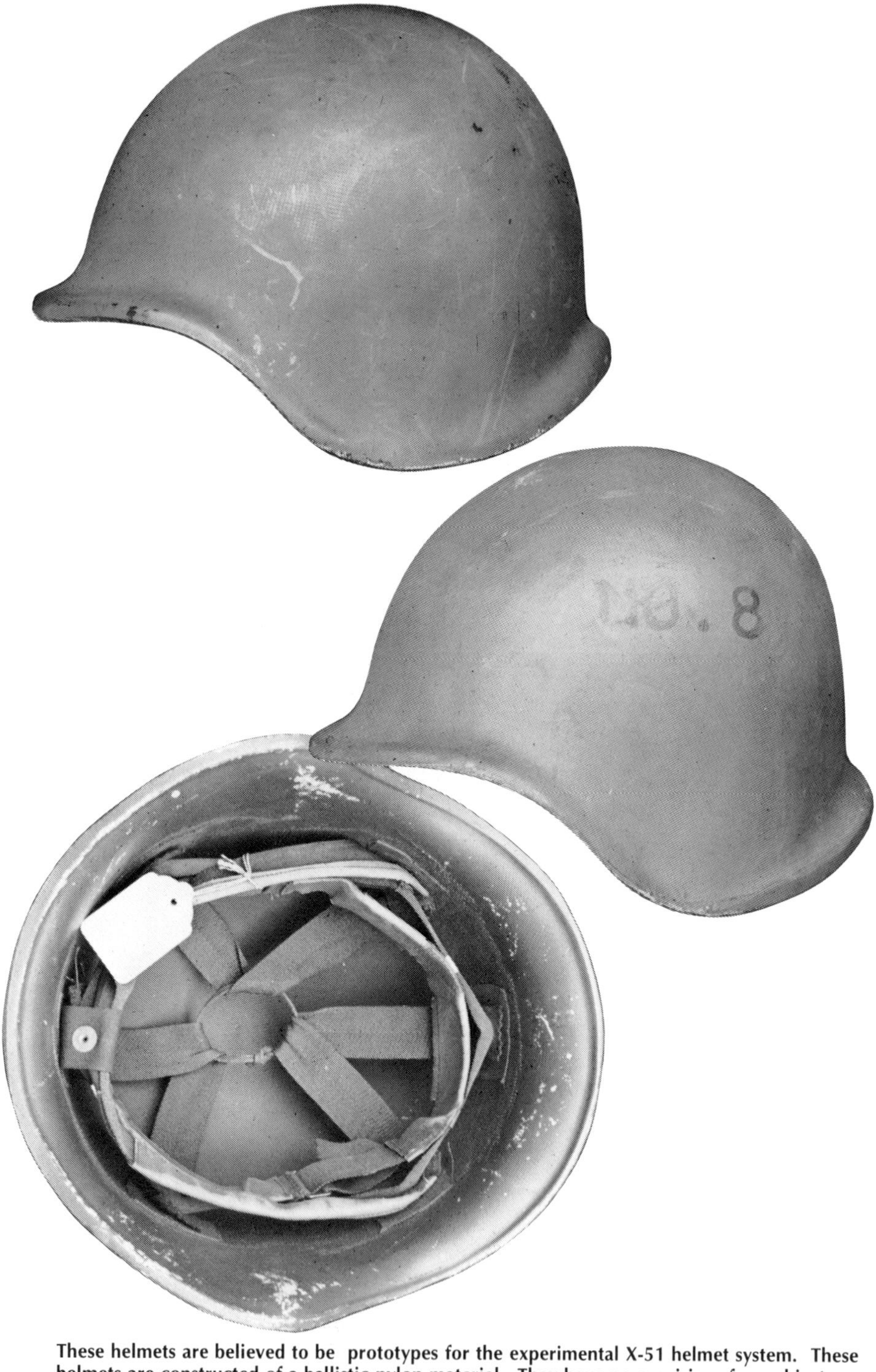

These helmets are believed to be prototypes for the experimental X-51 helmet system. These helmets are constructed of a ballistic nylon material. They have no provisions for a chinstrap. The suspension system was built directly into the nylon shell and has a slightly different configuration than a standard M1 helmet liner. The tag inside the helmet reads T21E2 Nylon Helmet. (Courtesy of Bill Rizzutti)

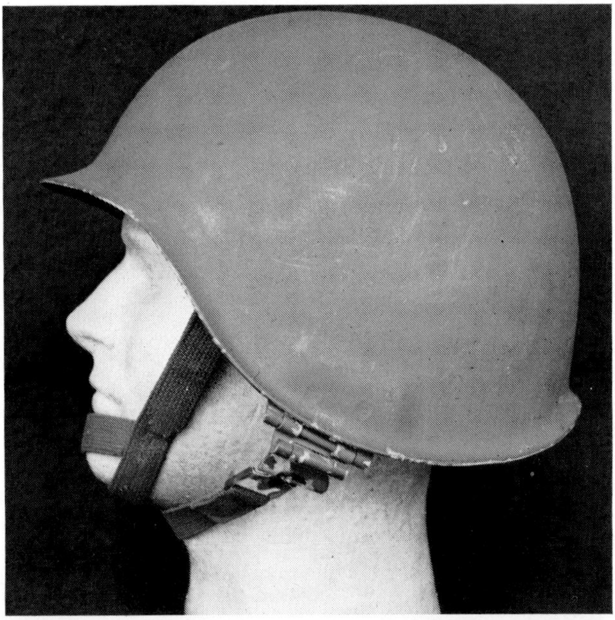

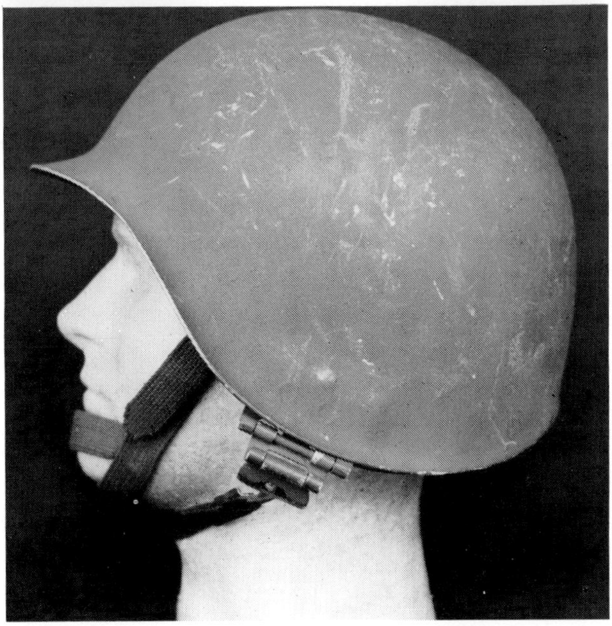

These helmets are two different versions of the experimental X-51 helmet. This helmet was constructed of aluminum. The helmet used a much larger and more complex system to hold the chinstrap in place. The chinstrap cup positioned over the chin is attached to the X-51 liner. The difference in these two helmets is most pronounced at the bottom rear of the shell. (Courtesy of Bill Rizzutti)

The X-51 was designed by Stephen Grancsay, the assistant to Bashford Dean at the Metropolitan Museum.

The Experimental Model X-51 being worn by an unknown general officer. The X-51 was made from an aluminum alloy. The number of X-51 helmets made is not known.

A U.S. Army photo showing an experimental Model X-49 made from ballistic nylon. Several versions of the X-49, X-51 style helmets were made from ballistic nylon. Notice the adaptation of the airborne "A" frame and leather chinstrap. (U.S. Army Photo Courtesy of the Quartermaster Museum)

Two different configurations of liners for the aluminum X-51 helmet system. Both of these liners use a neck band system which replaced the built-in neck band as used on the M1 helmet liner. (Courtesy of Bill Rizzutti)

This is the final variant of the aluminum-shelled combat helmets, the experimental X-53-2. This helmet used a large shell which fit over a liner made from ballistic nylon. The X-53 signaled a return to the more conventional type of helmet chinstrap loops and used the T-1 helmet clamps to hold the straps in place. (Courtesy of Bill Rizzutti)

The ballistic nylon liner system of the X-53 helmet. It seems apparent that the development of the X-53 nylon liner had a great influence on the modifications to the M1 helmet liner. The interior of the X-53 liner is one of the earliest known examples of the Vietnam era webbing configuration rather than the standard hammock style used in WWII and Korea. The shape of the PASGT is also starting to emerge. (Courtesy of Bill Rizzutti)

A closer look at the experimental 1968 LINCLOE Titanium helmet shell. Notice there is no rim seam on the helmet. It is painted a dark olive drab. The material used for texture appears to be silica gel.

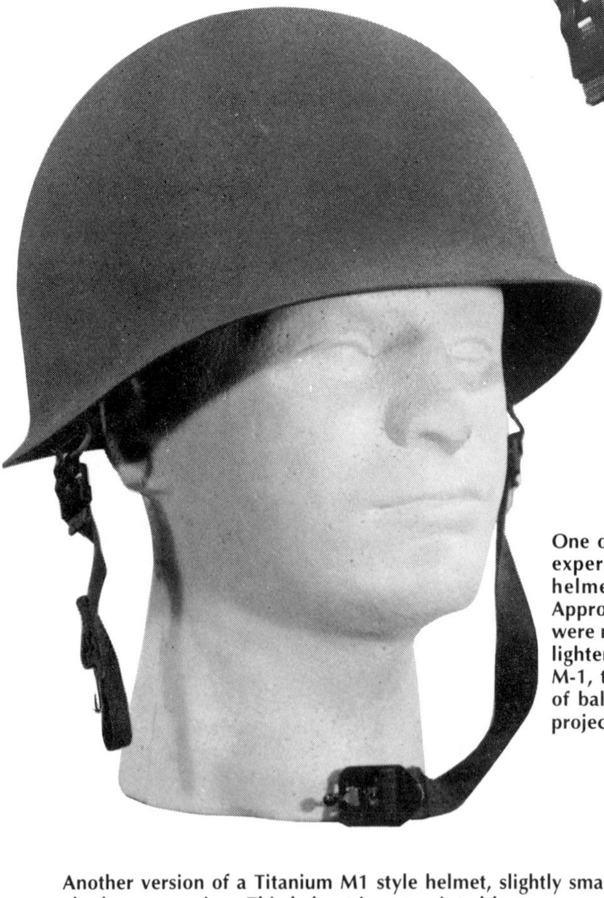

One of a few surviving examples of an experimental LINCLOE Titanium M1 helmet shell developed in 1968. Approximately 1,000 of these helmets were made for testing. While they were lighter and performed better than the M-1, they still did not correct problems of balance and fit. Consequently this project was shelved.

Another version of a Titanium M1 style helmet, slightly smaller than a standard M1 shell. It also has a raw rim. This helmet is not painted but appears to have been chemically treated producing the grayish-green color. The helmet is very smooth. The suspension is clipped directly into the titanium shell. (Courtesy of Bill Rizzutti)

In the later stages of the Vietnam war, secret experiments with a variety of materials were carried out at Aberdeen Proving Ground Maryland. These 1968 tests, compared the steel M1 with helmets made from nylon, titanium and titanium type III. During the test, the standard steel M1 shell fared the worst while the titanium shells provided the best protection. Little came from the experiment, only a few prototype titanium shells were produced but none apparently ever entered service. These tests did, however, signal a serious renewed effort to improve helmet design and protective ability.

In 1969, newspaper reports indicated a new helmet, radically different from the M1 and weighing half as much, was being developed by the Army. Beyond tantalizing bits of information such as this, little else is known.

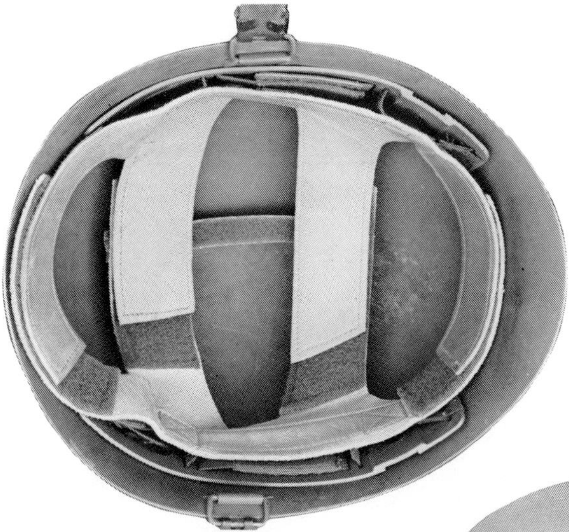

This is an example of the Welson-Davis style liner and is fitted to the interior of the Titanium style M1 helmet used in the LINCLOE tests. (Courtesy of Bill Rizzutti)

An M1-style helmet constructed of a ballistic material. Possibly a variant of the 1968 Nylon LINCLOE. The liner system uses an interesting "A" frame chinstrap assembly. (Courtesy of the David B. Powers Collection)

Prior to creating a new design, upgrading the material used in the M1 helmet was considered. This example, from the 1970s, is constructed of a ballistic nylon material. The new chinstrap assembly was tested by the Army in the 1970s. (Courtesy of the David B. Powers Collection)

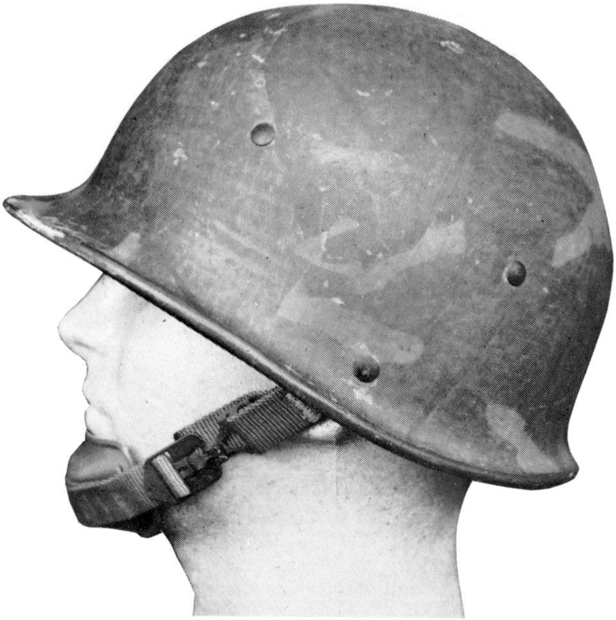

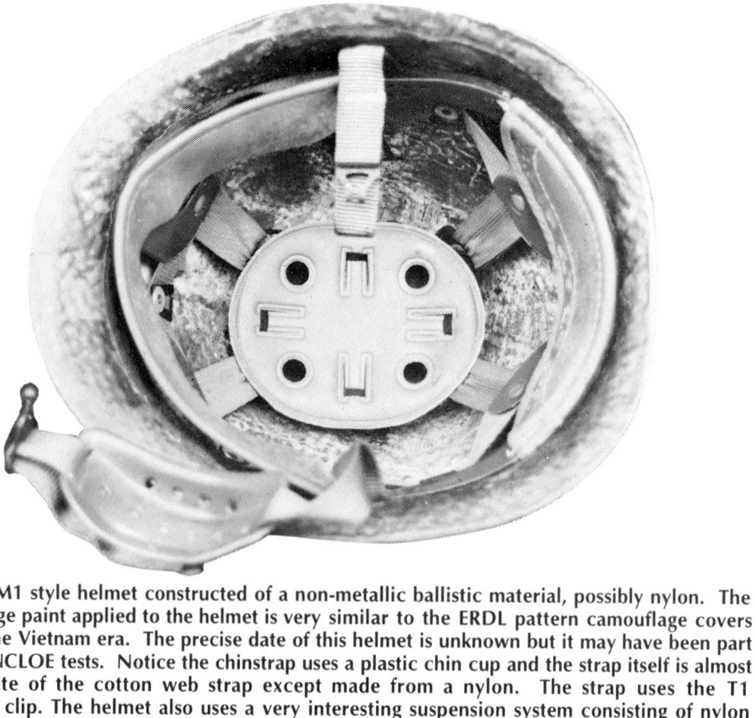

This is a M1 style helmet constructed of a non-metallic ballistic material, possibly nylon. The camouflage paint applied to the helmet is very similar to the ERDL pattern camouflage covers used in the Vietnam era. The precise date of this helmet is unknown but it may have been part of the LINCLOE tests. Notice the chinstrap uses a plastic chin cup and the strap itself is almost a duplicate of the cotton web strap except made from a nylon. The strap uses the T1 chinstrap clip. The helmet also uses a very interesting suspension system consisting of nylon and plastic. (Courtesy of Bill Rizzutti)

The task to actually develop the helmet fell to a little-known former Army Quartermaster Corps Laboratory called Natick, plus the Army Mechanics and Materials Research Center, (AMMRC) both located at the Watertown Arsenal in Watertown, Massachusetts. Soldiers' protection and opinions were being carefully studied. Initially, the Army set goals of improved protection and reduced weight. The Army hoped a lighter helmet would be more comfortable, and consequently, be worn with more regularity in combat situations. Intense scientific and medical studies were carried out on a variety of designs and materials. Steel, titanium, nylon, fiberglass and ceramics, in addition to various polymers and composite materials, were carefully evaluated. The winner was a fiber invented by Du Pont Chemical Company called "Kevlar."

Kevlar is a man-made fiber, and like the duckcloth used in making World War II M1 helmet liners, it was overlaid with a resin. Unlike a steel helmet, which was susceptible to dents, rusting or fragmenting, Kevlar would not be susceptible to such weaknesses. The material absorbed shock well and gave excellent protection against explosive fragments.

Designed by Philip Durand and Lawrence McManus, the helmet was touted as providing 25 to 40 percent greater protection than the M1. Its appearance was also very different from the M1. The Kevlar shell was produced in four sizes, and incorporated a small visor, covering the front of the head, the temple, ears and the side of the head. This "visor" added an additional 11 percent coverage. Like the M1, the Kevlar helmet shell was painted a dull olive drab.

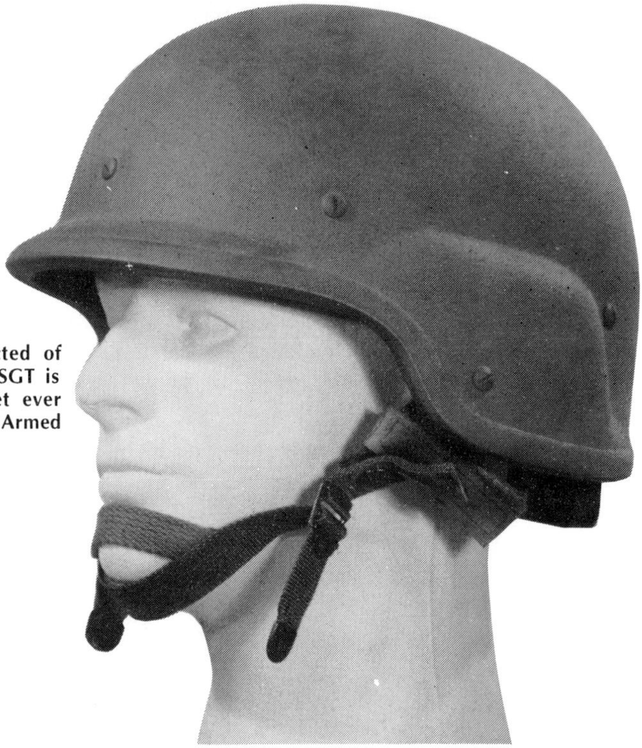

The PASGT. Constructed of molded Kevlar, the PASGT is by far the best helmet ever developed for the U.S. Armed Forces.

A very early example of one of the first molded prototype PASGT. This prototype, one of 36 made for testing, was constructed of an epoxy resin of the same density and weight of Kevlar. A key element of this helmet prototype is the use of the late-model M1 helmet chinstrap in nylon and the very squared-off appearance of the ear bulge. (Courtesy of Bill Rizzutti)

The Kevlar shell also incorporated a built-in suspension system. Gone was the need for a separate liner system as used in the M1. Comprised of nylon bands, a buckle was used to adjust the headband and a hook-and-pile system pull-tab was used to make drawstring height adjustments. The chinstrap, initially constructed of cotton but later changed to nylon, used a two-point suspension chin cup with dual adjustment buckles for maintaining accurate fit. The chinstrap was secured and released by using a single pull-the-dot snap fastener on the left side.

The dimensions for a medium sized helmet were 10.5 inches long, 9.5 inches wide, and 6.16 inches deep. Calculations showed this would fit some 50 percent of soldiers. Weight ran from 46 oz. for an extra-small helmet to 53 oz. for a large.

During the course of field tests, the PASGT (Personnel Armor System, Group Troops) was made in two weights. Approximately 3 pounds, and one and one-half pounds. The helmet's design and correct fit made it difficult for soldiers to tell the difference in the weight. Thus the decision was made to produce the new helmet in the three-pound weight because the heavier helmet had better protective qualities.

The helmet was officially designated the Helmet, Personnel Armor System Ground Troops, or PASGT, on 26 June 1978. As of May 1984 there were two manufacturers of PASGTs: Devils Lake Sioux Manufacturing Corporation and Federal Prisons Industries.

Another test model of the PASGT made from epoxy. This is probably one of the later prototypes as it has many of the features found in production model PASGT helmets. Early PASGT prototypes were tested with a variety of camouflage patterns. (Courtesy of Bill Rizzutti)

Externally, the differences between many of these helmets seem minor. However, the materials used in developing the PASGT were quite varied. This example is known as a G.R.P. model. It was made of fiberglass, reinforced with plastic. The fiberglass provided the ballistic protection while the plastic gave the helmet the rigid form it required to maintain its shape. G.R.P. was used for all the ballistic prototypes of the PASGT. Obviously one of the helmets has not been smoothed out and the fiberglass weave is very visible. (Courtesy of Bill Rizzutti)

This unfinished PASGT is constructed of a material described as "Kevlar felt." Unlike standard PASGT helmets made from regular woven Kevlar fibers, the material used in this helmet appears to have been pressed into the form of the helmet rather then built up, layer upon layer. (Courtesy of Bill Rizzutti)

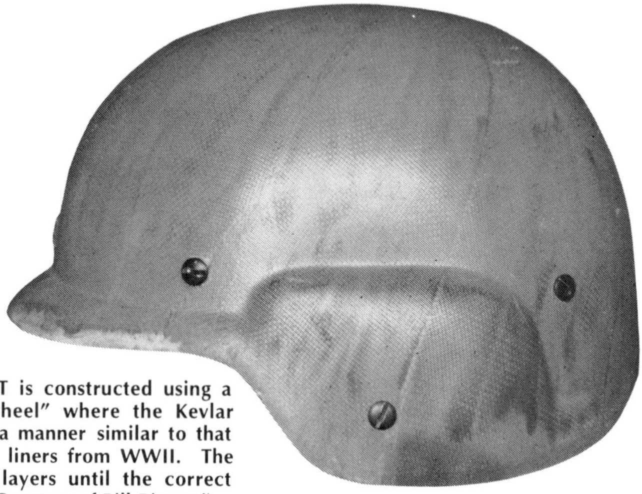

This unfinished PASGT is constructed using a method called "Pinwheel" where the Kevlar material is applied in a manner similar to that used on Westinghouse liners from WWII. The strips are applied in layers until the correct thickness is obtained. (Courtesy of Bill Rizzutti)

This PASGT was constructed using a method called "Squares." Very simply, approximately 5" by 5" squares of Kevlar were layered on top of each other until the desired thickness was obtained. (Courtesy of Bill Rizzutti)

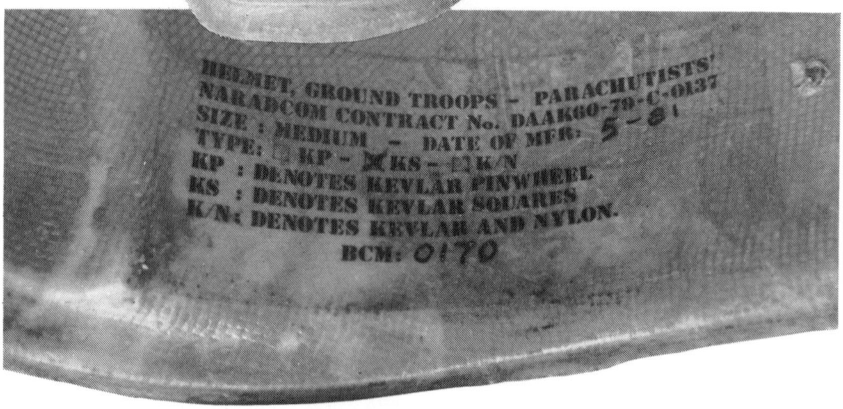

A nomenclature tag inside an unfinished PASGT. The tag identifies this helmet as being made using the Kevlar Squares method. Note that the contract number is dated 1979, and that the actual manufacturing date is May 1981. (Courtesy of Bill Rizzutti)

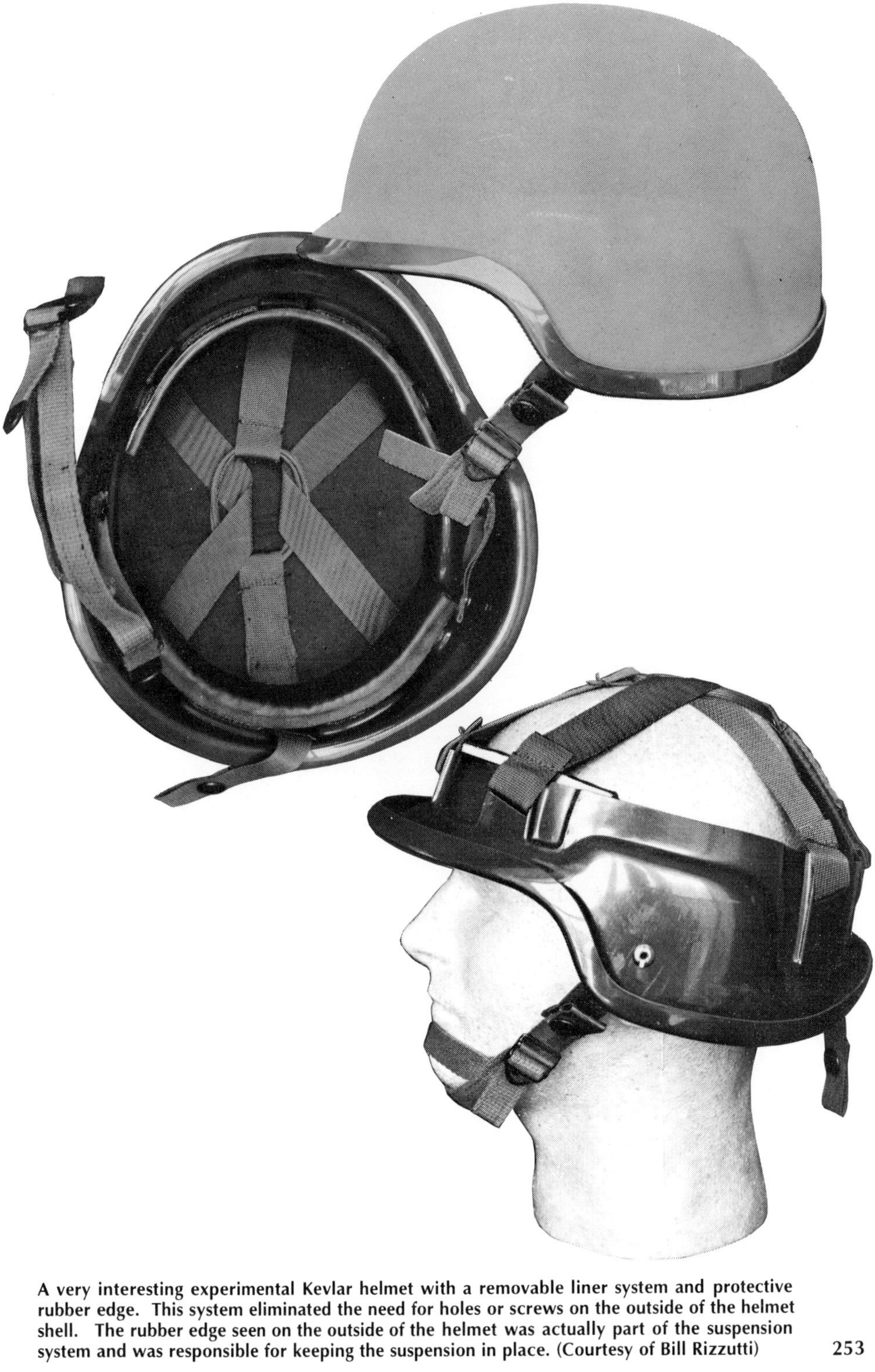

A very interesting experimental Kevlar helmet with a removable liner system and protective rubber edge. This system eliminated the need for holes or screws on the outside of the helmet shell. The rubber edge seen on the outside of the helmet was actually part of the suspension system and was responsible for keeping the suspension in place. (Courtesy of Bill Rizzutti)

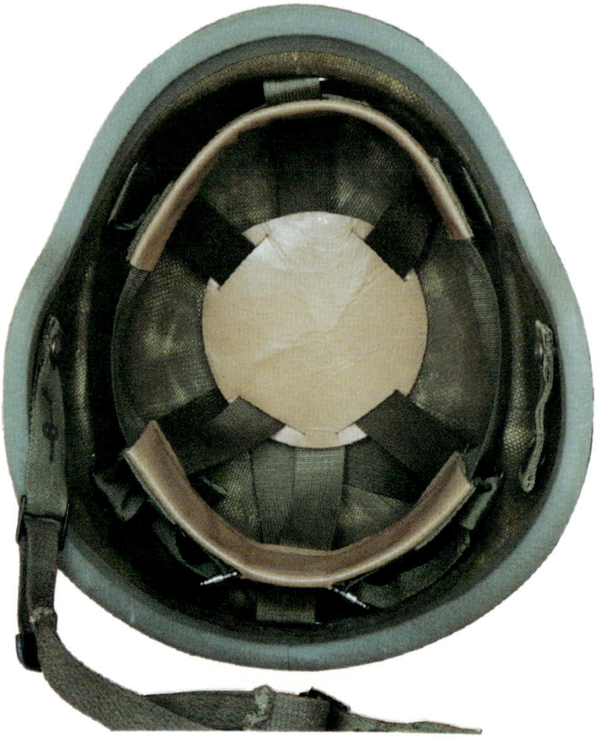

A PASGT helmet with an alternative liner. This liner was not adopted for use and was not widely produced. (Courtesy of Bill Rizzutti)

Except for the camouflage, after years of testing this is essentially what the production PASGT would look like. None of the camouflage paint jobs seen on the PASGTs presented in this book was adopted for production use. (Courtesy of Bill Rizzutti)

At some angles the PASGT resembles the German helmet of World War II. It is also interesting to consider how similar the PASGT is to the Experimental Model 5A. (See Chapter Five) One of the reasons the Model 5A was rejected was its similarity to the German M-1916 helmet World War I. While this similarity was initially noted, the PASGT protective qualities quieted the comparisons.

The PASGT covered approximately 11 percent more of the head than the M-1 helmet. The screw-heads visible around the helmet secure the suspension in place.

Looking inside the PASGT: The headband used inside the PASGT is nearly identical to the headband used on late-model M-1 helmets. As with the M-1 headband, it clipped into the suspension and used a buckle for adjustment. The nylon string at the crown was attached to a hook-and-pile (Velcro) tab which could be adjusted to improve the fit.

The internal suspension of the PASGT comes in three sizes: large, medium and small. In this photograph, the hook-and-pile (Velcro) connector has been lifted. The further down towards the back the tab was placed, the tighter the fit of the suspension. The suspension could be easily removed and cleaned by removing six screws from the helmet shell.

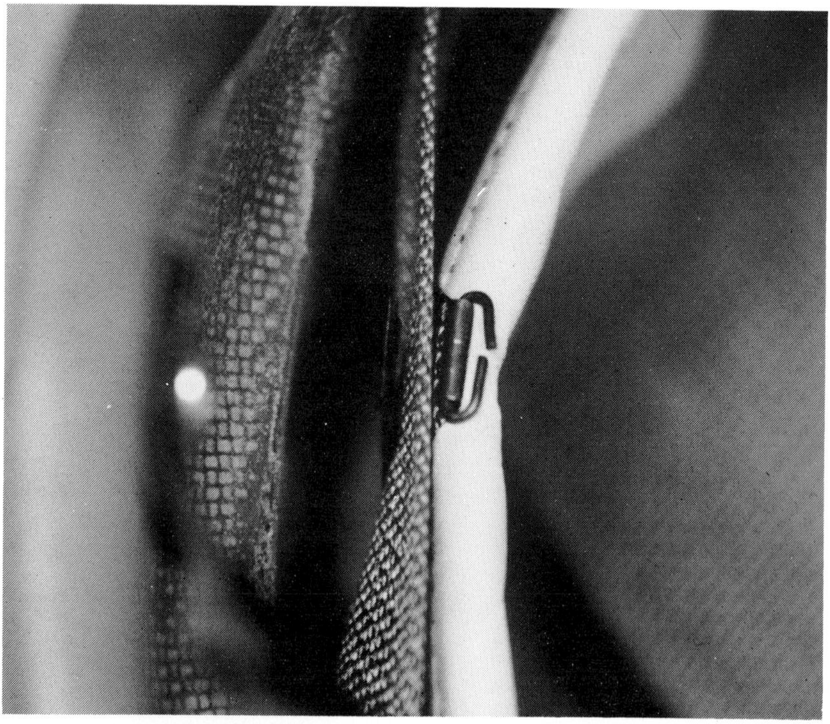

A close-up view of the clip system used to hold the headband in place.

Initially the U.S. Army ordered 1,000,000 PASGTs at a cost of approximately $97.00 per helmet. To date, more than 6,500,000 PASGTs have been manufactured by the four contracted makers. They include:

- Gentex Corporation of Carbondale, Pennsylvania (The first PASGT manufacturer)
- Specialty Plastics Products, Scranton, Pennsylvania
- Unicor Corporation, a Division of Federal Prison Industries, Bastrop, Texas
- Sioux Manufacturing, Fort Totten, North Dakota

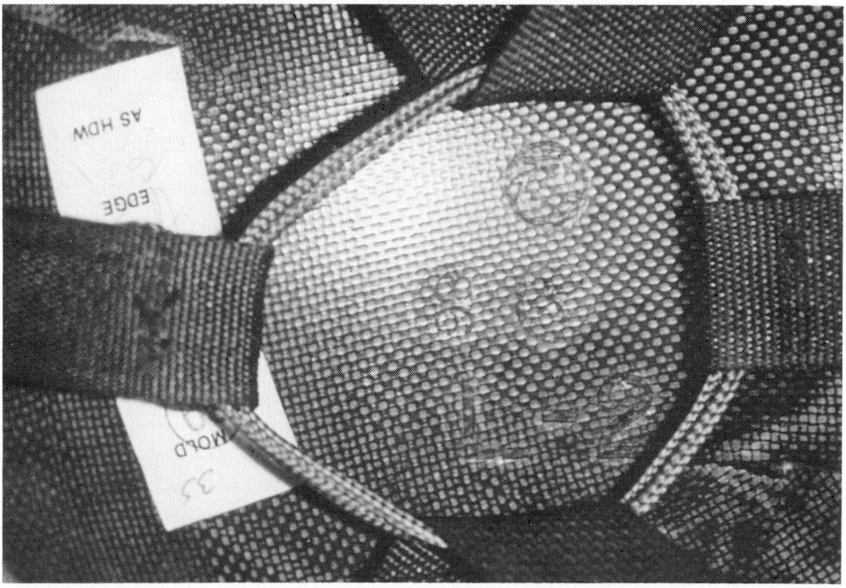

A close-up look at the interior crown of the PASGT. This is where each manufacturer placed the logo or makers marks. To date, four companies have made the PASGT. Besides makers marks, normally the helmet size and the year of manufacture are also located in the crown. The white paper label in this example is a manufacturer's quality control inspection sticker.

PASGT MAKER MARKS

Inside the crown

UNICOR

Gentex

SPP

Specialty Plastics Products

Sioux Manufacturing

In addition to these marks, each manufacturer was required to stamp in ink the following information inside each helmet shell: PASGT HELMET
DLA 100-85-C-456
NSN 8470-010092-7526
DEVILS LAKE SIOUX MFG
(Obviously DLA contract numbers and manufacturing firm names would be different.)

PASGT's are required to bear the maker's marks in the crown, plus display an ink stamping showing the maker's identity, date of manufacture and nomenclature labels.

"When did American soldiers start looking like the Waffen-SS" was a typical media comment when the PASGT was unveiled in combat during the 1983 Grenada invasion. While small features about the new helmet had appeared in the press on and off since 1976, the Grenada operation, dubbed "Urgent Fury," exposed many Americans to the new helmet for the first time. Initially, the press placed a great deal of emphasis on the profile of the helmet, which resembled the World War II German M1935. Within days of the invasion however, the helmet's outstanding protective qualities were confirmed when an 82nd Airborne paratrooper was shot in the head at a distance of 25 yards from a round fired from an AK-47. The bullet failed to penetrate the helmet and the soldier survived. Comparisons and complaints about the German helmet's resemblance soon faded away.

When issued, each helmet was intended to be hand fitted to the soldier and each helmet was issued with camouflage cover. The covers, while having the same camouflage pattern as those issued with the M1 helmet, were constructed differently. Covers from M1 helmets could not be used on Kevlar helmets or vice-versa. Along with the cover came a small instruction pamphlet. The pamphlet, entitled "This is your Ballistic Helmet," explained how to get a proper fit and described the helmet's nomenclature. The inside pack cover of the pamphlet was a "record of hits." This small questionnaire asked such questions as: Check what you were hit by: Grenade, mortar, artillery. Date of hit, location on helmet. The other side of the back cover was a pre-addressed post card which the soldier could drop in a mailbox and send to Natick for evaluation.

The PASGT consisted of primarily four parts:

- helmet shell
- suspension band
- headband
- chinstrap.

Each PASGT is issued with a camouflage helmet cover. Currently the patterns include Woodland, winter camo, and two desert schemes. PASGT covers were specifically designed for the helmet.

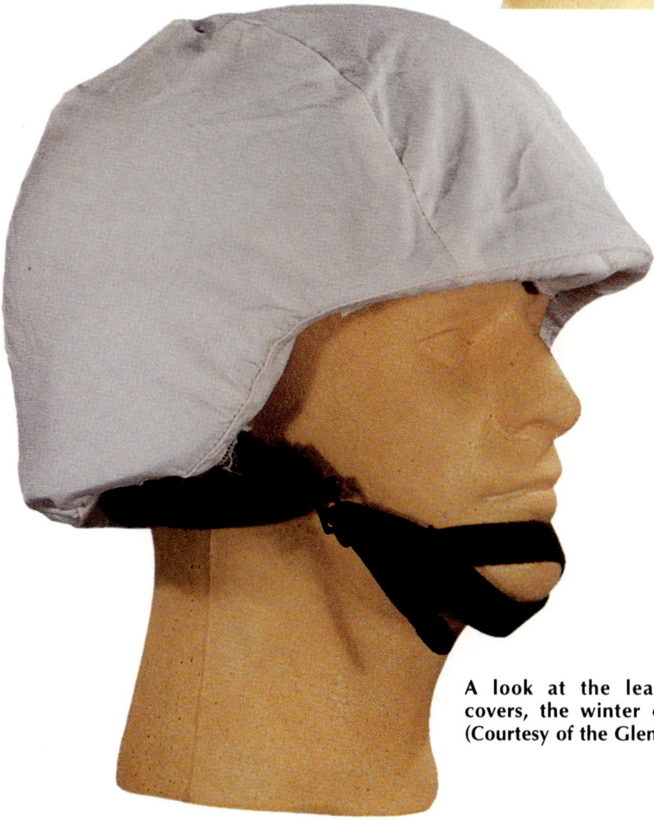

A look at the least common PASGT covers, the winter camouflage pattern. (Courtesy of the Glenn Haley Collection)

Typical contractors markings found on the interior of the PASGT camouflage cover. DLA stands for "Defense Logistics Agency" The 85 in the DLA number string indicates the year the helmet cover was produced. (Courtesy of the Glen Haley Collection)

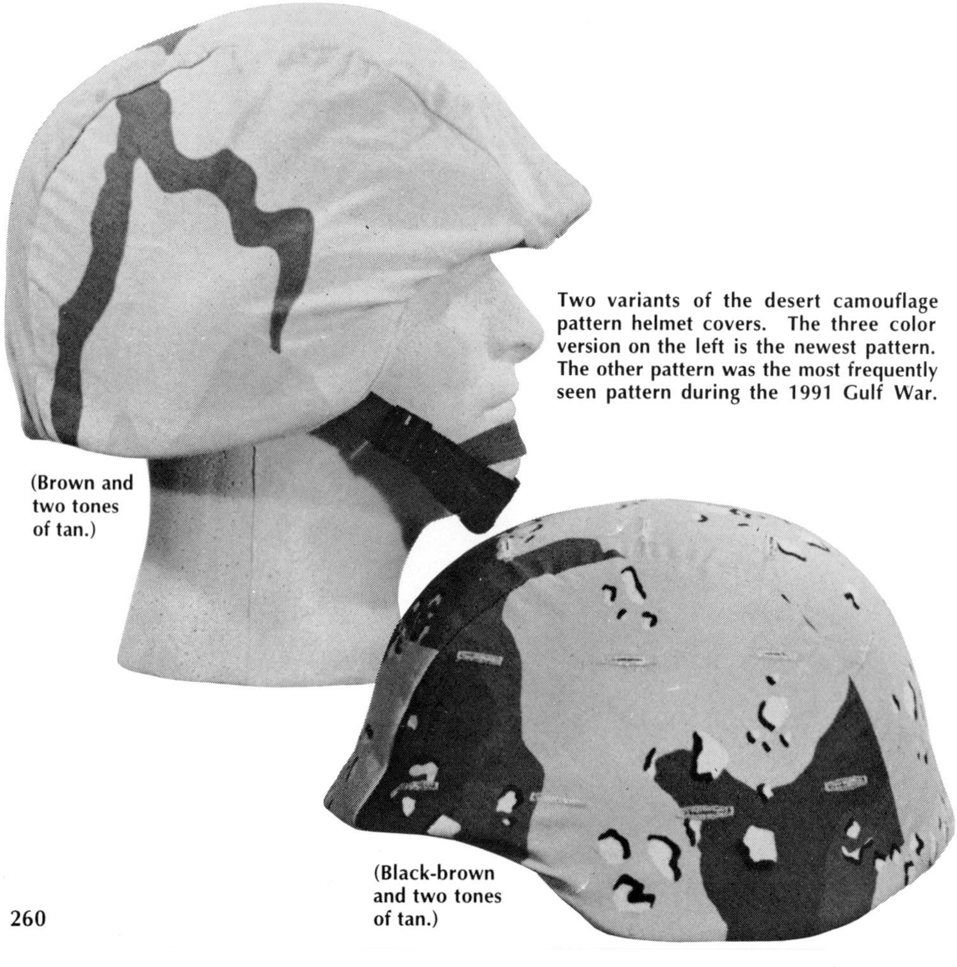

Two variants of the desert camouflage pattern helmet covers. The three color version on the left is the newest pattern. The other pattern was the most frequently seen pattern during the 1991 Gulf War.

(Brown and two tones of tan.)

(Black-brown and two tones of tan.)

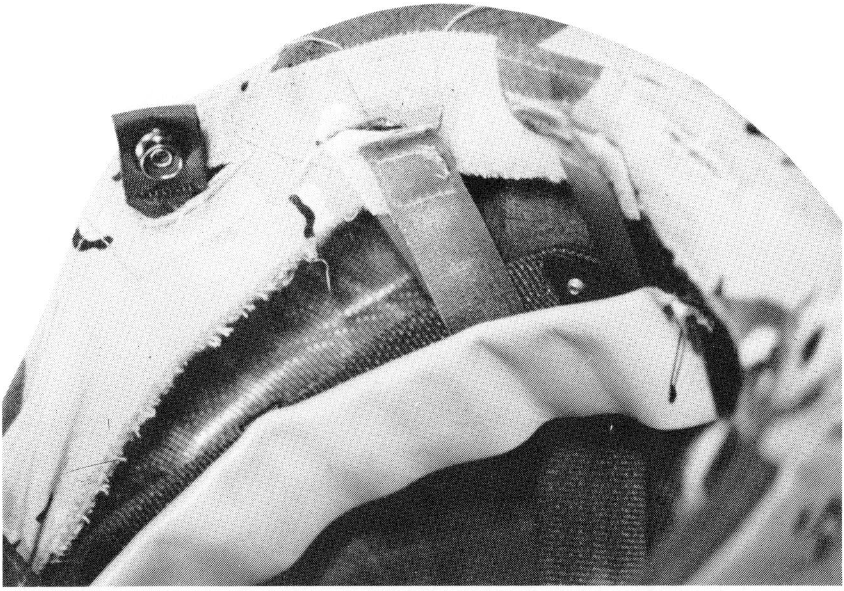

The most complicated operation for most users of the PASGT is installing the camouflage cover. Like the helmet and the suspensions, they are sized. The Velcro tabs slide beneath and around the suspension to hold the cover in place. Slits in the cover allow chinstraps and the paratrooper retention straps to pass through.

The helmet's suspension band resembled the internal suspension of the M1 liner in that it retained the hammock-style appearance. However, the suspension was adjustable by using a hook-and-pile system. The headband was nearly identical to the M1 headband. It retained the leather headband and the metal clips which snapped around the suspension band. The headband was equipped with an adjustment buckle. The new internal suspension also lowered the height of the helmet on the head. While the M1 helmet sat an average of two inches above the head, the new PASGT rested a mere half-inch above.

Six screws held the suspension system in place. These could easily be removed to facilitate cleaning the inside of the helmet or replacing the suspension system.

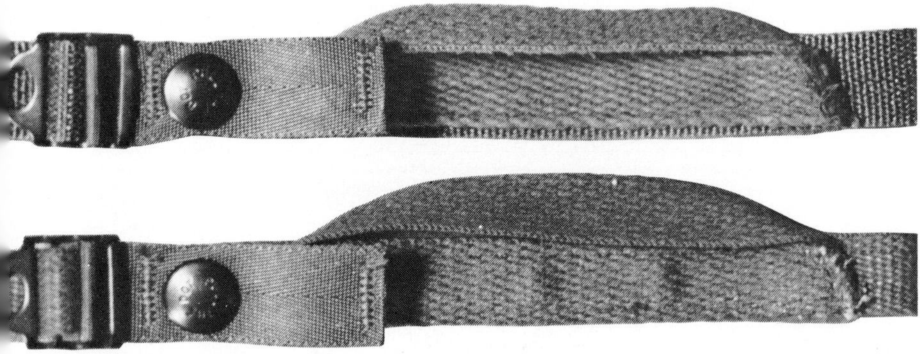

The chinstrap of the PASGT. The bottom chinstrap is the early version made from cotton while the top strap is constructed of nylon. The actual chin-cup areas were cotton on both versions.

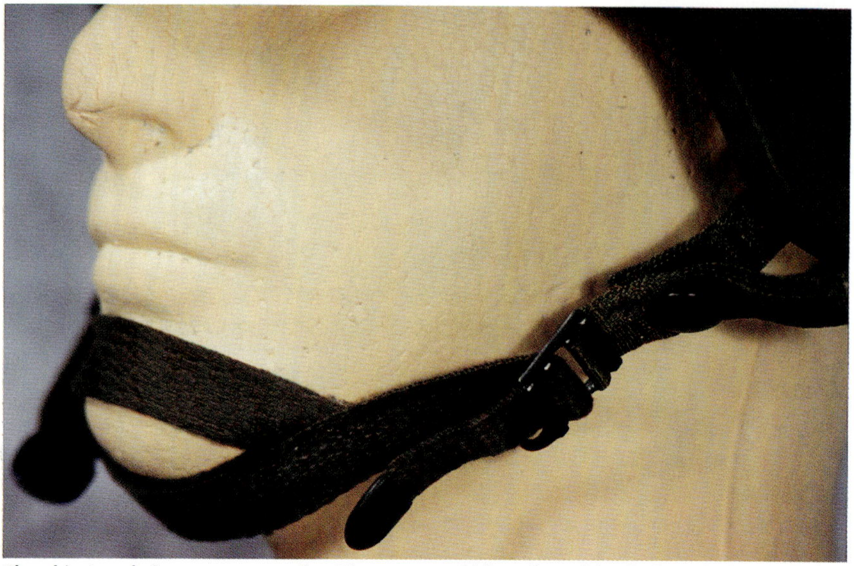

The chinstrap being worn correctly. The strap could be adjusted for fit from both sides. The chinstrap release is on the left side. A sharp tug disconnects a press-the-dot fastener.

The nylon chinstrap ends connected to the helmet shell are held in place by stainless steel screws.

The PASGT could also be converted to Airborne use with the addition of an Ensolite foam impact pad and special parachutist retention straps. To install the pad and the straps properly, a slightly larger screw (5/8") replaced the standard screw at the rear of the helmet. The retention straps were threaded through slits placed in the helmet cover and then hooked around the helmet's chinstrap.

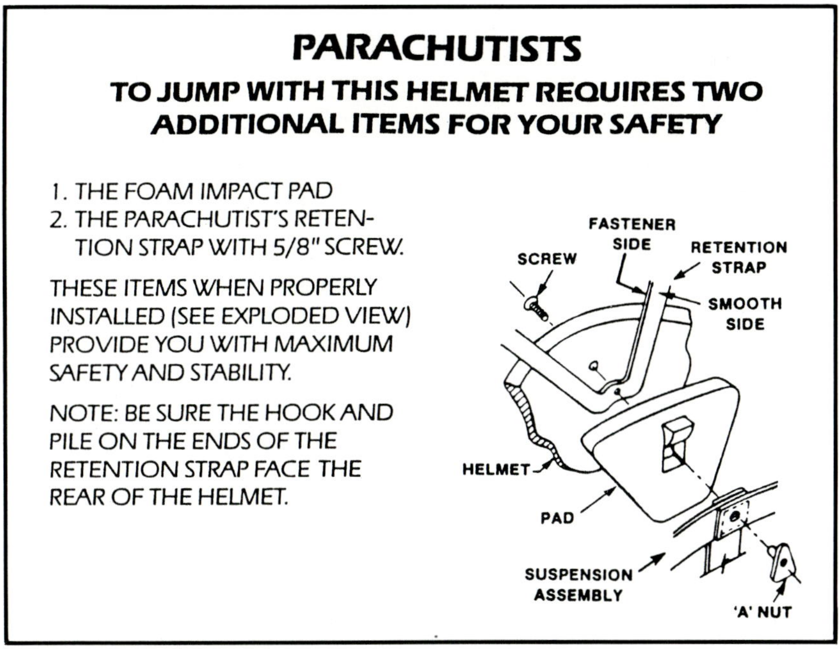

PARACHUTISTS

TO JUMP WITH THIS HELMET REQUIRES TWO ADDITIONAL ITEMS FOR YOUR SAFETY

1. THE FOAM IMPACT PAD
2. THE PARACHUTIST'S RETENTION STRAP WITH 5/8" SCREW.

THESE ITEMS WHEN PROPERLY INSTALLED (SEE EXPLODED VIEW) PROVIDE YOU WITH MAXIMUM SAFETY AND STABILITY.

NOTE: BE SURE THE HOOK AND PILE ON THE ENDS OF THE RETENTION STRAP FACE THE REAR OF THE HELMET.

A diagram showing the proper installation of the parts required to convert a standard PASGT to the airborne configuration

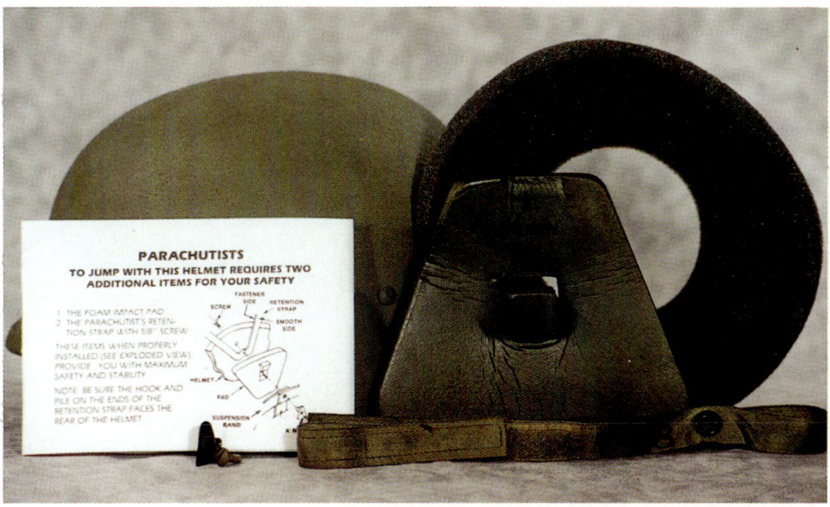

The PASGT can easily be converted to airborne use by adding three components: An ensolite impact pad, airborne retention straps, and a longer screw to hold the items in place. The "donut hole" pad for the crown is an optional item.

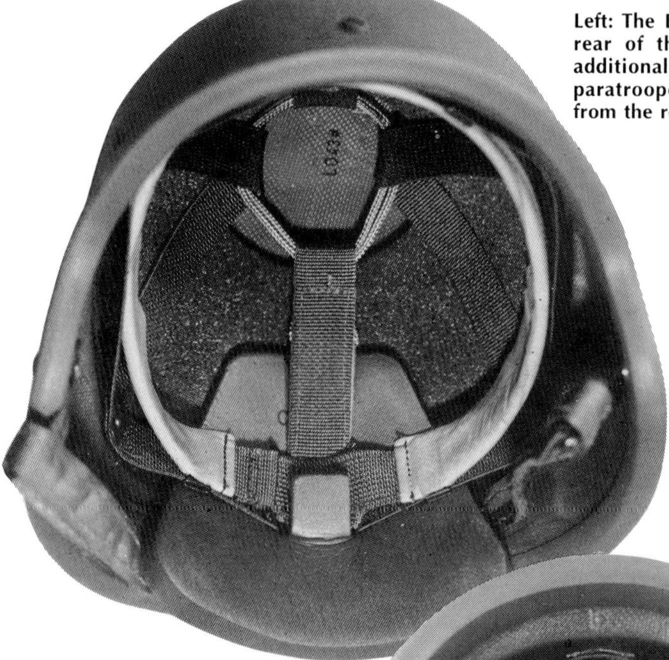

Left: The Ensolite pad rests against the rear of the helmet shell to provide additional padding. Again, note the paratrooper retention straps coming from the rear of the helmet.

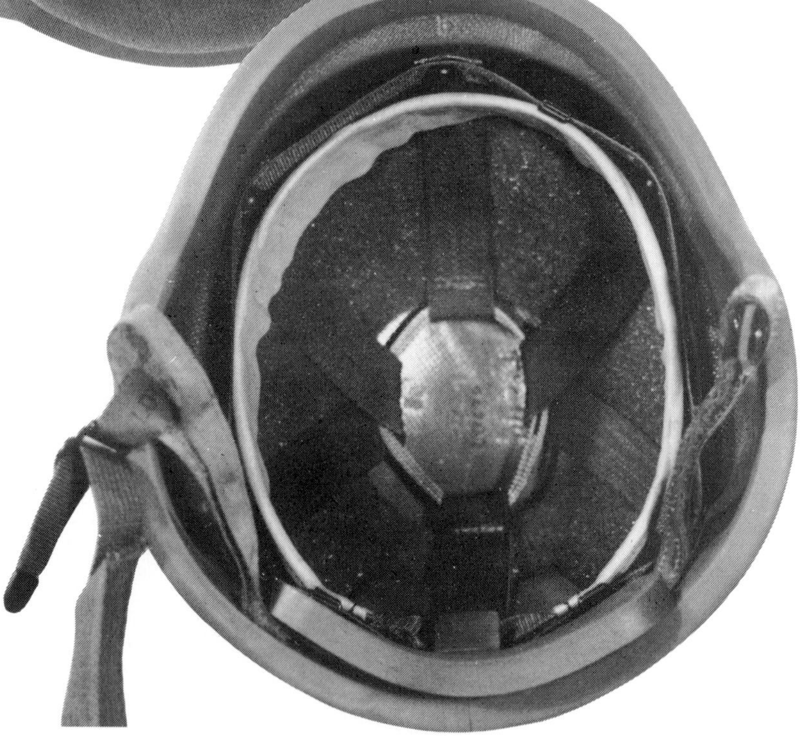

Below: A PASGT in its airborne configuration. Notice the paratrooper retention straps run from the rear of the helmet to the chinstraps. They help to secure the helmet in place during jumps. Again, the gray foam pad in the crown is an optional item which can be purchased by individual GI's.

The PASGT is proof of the Army Research Lab's commitment to outfitting American Armed Forces personnel with the best equipment available. Recognized as the finest protective helmet in the world, the PASGT is used by thousands of American military personnel around the world. Many other countries have adopted versions of the helmet as well.

SSgt Scott Elliott, a member of the 314th Airlift Wing Public Affairs Office, wears the PASGT while deployed to Tusla Airfield, Bosnia. Besides the Kevlar helmet, he is also equipped with the other part of the Personal Armor System, the flak vest. Like many U.S. military members, SSgt Elliott has printed his name and rank on the neoprene band on his helmet. (Photo courtesy of Scott Elliott)

Wearing a PASGT a U.S. Army member gathers some information from a member of the Bosnian military. The helmet worn by the Bosnian soldier is a variation of the U.S. M1. Obviously the camouflage pattern is different from U.S. covers, plus the elastic band holding the helmet cover in place and the chinstrap are different. (Photo courtesy of Scott Elliott)

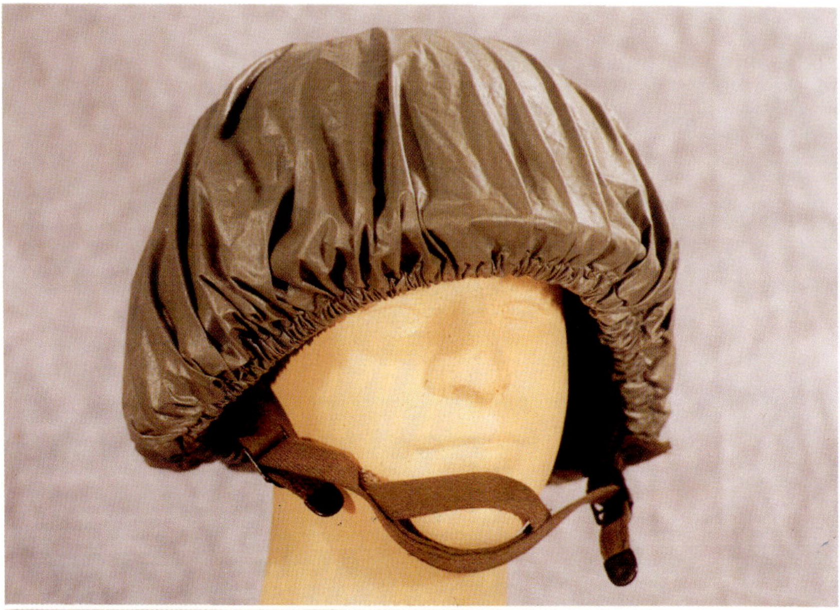

This obscure helmet cover is the Cover, Chemical Protective and was constructed of the same material used for the protective hoods on M-17 series gas masks. The cover could be worn on the PASGT or the M1.

An experimental Kevlar helmet made by Sierra Company in the 1980's. This interesting helmet and shield system was developed for use by crewmen in battleship turrets to protect them from the blast. (Courtesy of the Keith R. Jamieson, MD, Collection)

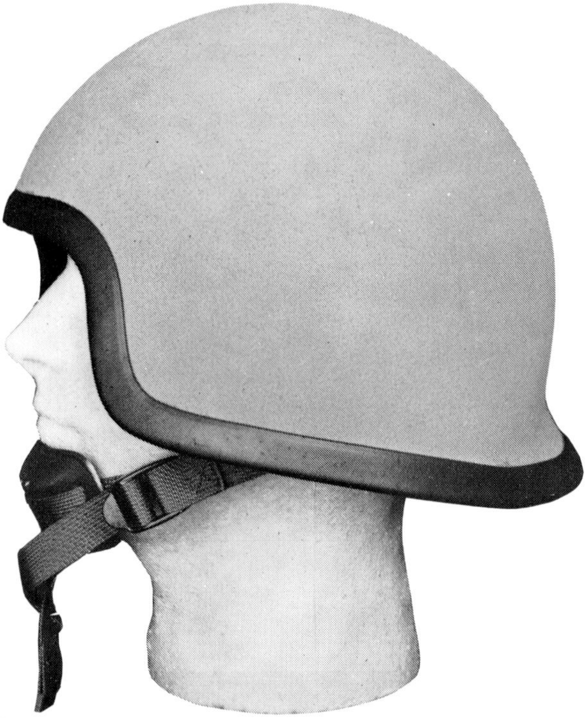

An updated version of the US Navy Talkers helmet identified by the tag as being "Helmet, Phonetalker MK4 Model O." The helmet is constructed of a very heavy ballistic nylon weave. The liner system consists of individual foam pads with adhesive backings attached to the inside of the helmet shell. This example is dated 1983. It seems most, if not all MK4 helmets were made by Bell Helmet Company. (Courtesy of Bill Rizzutti)

A M1 helmet liner used as a suspension system for a NBH-3 Navy helmet. The small black and white squares are adhesive-backed and stick inside the helmet liner. (Photo courtesy of David B. Powers)

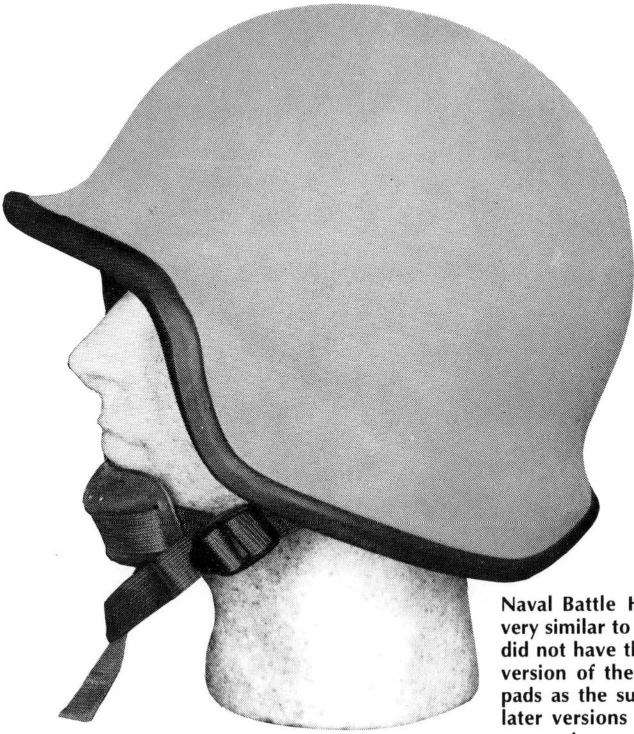

Naval Battle Helmet NBH-1. While very similar to the PASGT, the NBH-1 did not have the ear bulges. An early version of the NBH-1 used adhesive pads as the suspension system while later versions came with a built-in suspension system. (Courtesy of Bill Rizzutti)

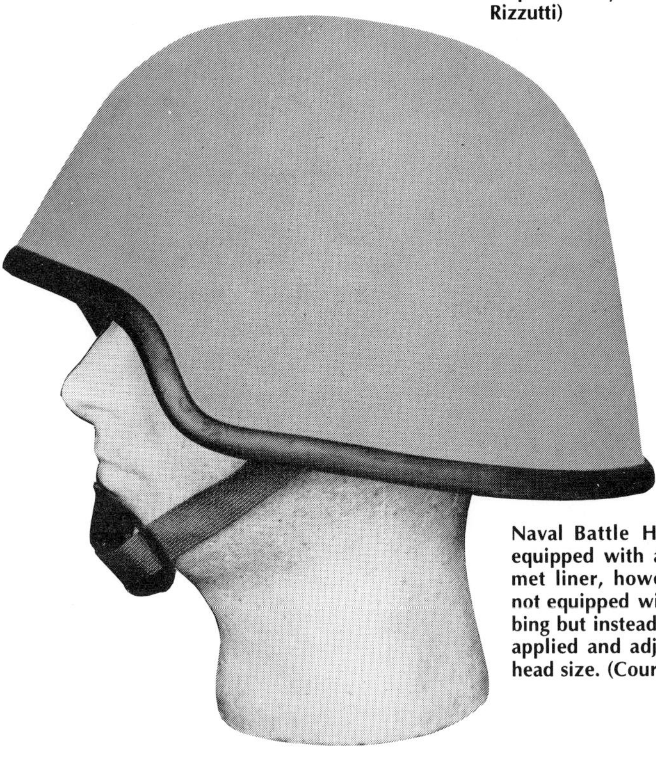

Naval Battle Helmet NBH-3 was equipped with a standard M1 helmet liner, however, the liner was not equipped with suspension webbing but instead used adhesive pads applied and adjusted according to head size. (Courtesy of Bill Rizzutti)

BIBLIOGRAPHY & REFERENCES

America's Munitions ....Benedict Crowell 1919

America's Tin Hat...Chris Armold 1996

An Estimate of the Military Value and Desirable Characteristics of Armor Helmets For Ground Forces. .... (Project ALCLAD) SECRET ....Norman H. Hitchman 1950

A Visor for the standard M1 Helmet to Prevent Injuries to the Face ....................U.S. Army Project X-227 1944

British Steel Helmets, 1915-1918.................Marcus Cotton 1995

Body Armor for Modern Battle ....Col Gene R. Studler 1946

Battlefronts of Industry..................................David Woodbury unk

Care and Use of Individual Clothing and Equipment ....U.S. Army FM 21-15 1977

Combat Flying Equipment ....C. G. Sweeting 1989

Combat Helmets of the World ....Paolo Marzetti 1996

Comfort With Protection Restricted ....Army Ordnance Report Unk

Contributions to Science and Technology ....Army Research Lab 1995

Development of a one piece infantry helmet ....Natick 1976

Fighting Man's New Headgear......................Major H.G. Sydenham 1941

Head to Toe Protection ....Nat Wood 1968

Heat Treating and Cold Rolling Hadfield Manganese Steel.....................................U S Patent Office 1943

Helmets and Body Armor in Modern Warfare ....Bashford Dean 1919

Helmets and Body Armor in Modern Warfare ....Stephen V. Grancsay 1945

History of the Helmet Liner ....Marion Massen 1944

Keep Your Helmet, Keep Your Head ....Norman Hitchman 1957

Military Helmet Design ....Frederick J. Lewis 1958

Nine Million Hadfield Manganese Steel Helmets ....Unattributed 1920

Project relating to Helmets and Body Armor ....Ordnance Dept 1945

Reference Manual for Personal Equipment Officers ....AAF 1945

Report of Test of Net, Helmet, Camouflage..QM Project T355 1944

Specification: Ball, Chin Strap Release ........U.S. Army A 7114390 1944

Specification: Buckle, Sheet Brass................U.S. Army A 139646A 1944

Specification, Chin Strap Release, T1 ....QM Project T376 1944

Specification: Clip, Brass .............................U.S. Army A 139648 1944

Specification: Drawing M1 Helmet and Straps	U.S. Army MIL-H-10990	1951
Specification: Drawing M1 Helmet and Straps	U.S. Army MIL-H-1988E	1966
Specification: Drawing M1 Helmet and Straps	U.S. Army MIL-H-41802B	1964
Specification: Helmet, M-1917A1 Assembly	U.S. Army D7002	1933
Specification: Helmet, M3	U.S. Army 202432	1943
Specification: Helmet, M4	U.S. Army 202433	1943
Specification: Helmet, M4A1	U.S. Army 202437	1944
Specification: Helmet, M4A2	U.S. Army 202439	1945
Specification: Helmet, M5	U.S. Army 202441	1944
Specification: Helmet, Soldiers, Steel M1	U.S. Army MIL-H-1988A	1956
Specification: Helmet, Steel, M1 Assembly and Details	U.S. Army D 2-1-87	1951
Specification: Helmet, Steel, M1 (without lining)	U.S. Army D39081	1941
Specification: Helmet, Steel, M1 With Lining Assembly	U.S. Army D39085	1941
Specification: Helmet, Steel, M1C (parachutists, without lining)	U.S. Army D7161936	1945
Specification: Helmet, Steel, Parachutists M2 (without lining)	U.S. Army D44085	1942
Specification: Helmet, Steel, Parachutists M1 (with lining)	U.S. Army D44083	1944
Specification: Hinged Loop Assembly	U.S. Army A 7160039	1945
Specification: Hook, Brass Wire	U.S. Army A 139649A	1944
Specification: Hook, Chin Strap Release	U.S. Army A 7114389	1944
Specification: Liner Helmet, (Parachutists) M1	U.S. Army CQD No 66	1942
Specification: Lining, Helmet, Assembly (M1)	U.S. Army D39082	1941
Specification: Lining, Helmet, Assembly (M1917A1)	U.S. Army D6863	1933
Specification: Lining, Helmet, Steel, Parachutists	U.S. Army D44083	1944
Specification: MK II Talkers Helmet	MK II (Unattributed)	1944
Specification: Reclamation of Helmet, Steel, M1	U.S. Army JQD No 2151	1946
Specification: Release, Chin Strap	U.S. Army A 71144405	1945
Specification: Strap, Chin, Ground Troops, Helmet, Steel, M1	U.S. Army MIL-S-43841A	1977
Specification: Strap, Chin, Helmet, Long, Assy	U.S. Army B 170329	1943

Specification: Strap, Chin, Helmet, Short, Assy	U.S. Army A 192228	1943
Specification: Strap, Chin, Webbing, Liner, M1 Parachutist's	U.S. Army CQD No 205	1944
Specification: Suspension, Chin, Details, Liner, Helmet M1	U.S. Army QM No 2-1-36	1944
Specification: Webbing, Cotton, Natural or in Colors	U.S. Army No. 6-185C	1941
Streamlined Headgear	Maj A.M. Brumbaugh	1942
Summary of Alloy Steels for Helmets and Armor	Unattributed	1918
The Bashford Dean Collection of Arms and Armor	Stephen Grancsay	Unk
The Steel Pot	Robert Southee	1982
The Tin Hat	Mark Reynosa	1996
Tin Hats	Gary I. Crokett	1940
The M1 Helmet, 1941-1945 A Short History	Mark Reynosa	1994
The New Combat Helmet	Col Rene R. Studler	1942
The New M1 Soldiers Helmet	Unattributed	Unk
The Ordnance Department Procurement and Supply	Thomson & Mayo	1960
The Story of the New American Helmet	Daniel L. Wells	1944
25 Years. The Association of Cotton Textile Merchants of New York	Unattributed	1944
U.S. Army Uniforms of the Vietnam War	Shelby Stanton	1989
Wanted: An American Helmet	James A. Reilly	1937
What About the Helmet	John R.D. Cleland	1954
What Tomorrow's GI's Will Wear	Rudy Arnold	1962
Wound Ballistics	U.S. Army	1962